Immortality Inc.

The Science and Business
of Living Forever

By
Arlo Voss

Immortality Inc.

The Science and Business of Living Forever

Contents

Introduction

The pursuit of longevity has accompanied humanity throughout history, reflecting our enduring quest for understanding and mastery over the natural world. The allure of extending life, potentially even to immortality, raises profound questions that span science, philosophy, economics, and ethics. This journey is as much about the mysteries of life as it is about confronting the limits of human existence.

In recent decades, the rapid pace of technological and scientific advancement has significantly transformed our approach to longevity. What was once the domain of myth and legend is gradually becoming a plausible scientific endeavor. The intersection of biology, technology, and society has brought about a renaissance in longevity research, propelling the dream of significantly extending human life towards reality. As we stand on the cusp of these discoveries, it becomes essential to navigate the complex tapestry of opportunities and challenges inherent in this pursuit.

Our contemporary exploration of human lifespan is driven by a myriad of scientific fields. From the burgeoning study of cellular regeneration to breakthroughs in genetic engineering, the foundational elements of life itself are being scrutinized and manipulated to unlock the secrets of aging. Various technological innovations, like artificial intelligence, further enhance our capability to analyze and predict aging processes, creating unprecedented opportunities for tailored health interventions (Kondrotas et al., 2020). Each step forward invites new ethical, social, and economic questions, demanding a holistic approach to the exploration of human longevity.

The philosophical dimensions of longevity are deeply intertwined with scientific exploration. Questions about the nature of existence, the value of life, and the implications of an extended lifespan challenge us to rethink fundamental beliefs. The potential for immortality poses questions about the essence of humanity itself. Understanding these philosophical perspectives provides a broader context, enriching the scientific inquiry with insights into what it means to live and to live longer, as well as the societal values we hold dear.

The implications of longevity extend beyond personal aspirations, touching on complex societal and economic dynamics. An increase in lifespan could reshape demographic patterns, necessitating reforms in healthcare, economics, and urban planning. The economic impact is profound, as industries adjust to cater to an aging yet active population. Evaluating these shifts requires a keen understanding of both current structures and the visionary policies needed to sustain societal well-being alongside technological progress.

Ethically, the pursuit of longevity confronts us with dilemmas of access and inequality. The possibility of unequal distribution of life-extending treatments could exacerbate existing social disparities. Additionally, the ethical management of biotechnology and genetic manipulation necessitates robust oversight to mitigate potential abuses. Navigating these challenges demands an inclusive dialogue that considers diverse perspectives and strives for equitable solutions (Caplan, 2019).

This book aims to provide a comprehensive exploration of these multifaceted issues. By integrating scientific, philosophical, and ethical discussions, it aspires to inform, educate, and inspire its readers. In doing so, readers will be encouraged to critically engage with the subject, considering the broader implications of humanity's potentially changing relationship with life and death.

Throughout the chapters, a wide array of topics will be explored—from the historical quest for immortality to the latest advancements in genetics and artificial intelligence. The implications of these developments will be examined through sociological, economic, and cultural lenses. The goal is to foster a nuanced understanding of longevity, preparing readers to contribute thoughtfully to the ongoing conversations and decisions shaping this transformative journey.

Ultimately, the focus is not merely on extending life but on enhancing the quality of life. As technological advancements offer new possibilities, they also require us to weigh choices carefully. Our collective responsibility is to pursue longevity in a manner that respects human dignity, promotes equality, and aligns with an ethical commitment to future generations. In embarking on this exploration, we honor the complexity of our shared human experience while daring what was once unthinkable.

In the end, the journey towards longevity is a testament to the depth of human curiosity and the persistent drive to expand the boundaries of our potential. It invites us to reflect on the profound connections between science, society, and individual aspiration. As we venture into this new frontier, our understanding of life itself may be transformed, inviting us to see longevity not just as a scientific endeavor but as an unfolding narrative of the human spirit.

Chapter 1:
The Quest for Immortality

In the labyrinthine corridors of human aspiration, the quest for immortality stands as one of the most profound and enduring pursuits. Our journey begins by tracing the ancient roots of humanity's desire to transcend the bonds of mortality, a quest driven less by arrogance and more by a deep-seated yearning to understand the universe and our place within it. Ancient myths and legends from diverse cultures have spun tales of elixirs and secret realms where death is but a fleeting shadow. Philosophers and alchemists, thinkers and dreamers alike, have pondered the implications of eternal life, raising questions that straddle the lines of science and metaphysics (Smith, 2003; Johnson & Lee, 2010). As we delve into the philosophical underpinnings of these pursuits, the ambition for extended life is tempered by ethical dilemmas and existential musings on what it means to live forever. This narrative sets the stage for an exploration of both the steam propulsion of modern science and the winds of age-old wisdom, each propelling the voyage towards the elusive shores of immortality.

The History of Human Longevity

The quest for a longer, healthier life has intrigued humans throughout history, shaping our cultures, philosophies, and scientific endeavors. This journey is characterized by a blend of myth, magic, and burgeoning scientific understanding. Ancient texts from various civilizations, such as the Epic of Gilgamesh and the Taoist writings from ancient China, have pondered immortality, underscoring humanity's deep-rooted desire to transcend mortality.

For centuries, people sought the elixir of life, a mythical potion granting eternal youth. Alchemists in medieval Europe were known for striving to discover this sorcerous substance, viewing it as the pinnacle of human achievement. Despite these early efforts being more fantastical than factual, they laid the groundwork for our modern understanding of longevity as something science could eventually decipher and manipulate.

Interestingly, practical advances in human longevity didn't materialize through magic but rather through incremental improvements in living conditions. The Enlightenment era brought about advancements in medicine and public health, significantly reducing mortality rates through vaccination, improved hygiene, and better nutrition (Riley, 2001). These changes meant people began to live longer, healthier lives, igniting newfound hope in the possibility of extended life spans.

The average life expectancy has increased significantly over the past few centuries. According to historians, life expectancy in ancient Rome was only about 30-35 years, primarily due to high infant mortality rates and the prevalence of diseases and harsh living conditions (Frier, 1994). However, with the advent of modern medicine and improved socioeconomic conditions, today's life expectancy in developed countries often surpasses 80 years.

One of the critical turning points in the quest for longevity was the discovery of antibiotics in the 20th century. These drugs heralded a new era wherein infectious diseases, once a dominant threat, were increasingly treatable (Podolsky, 2015). By curbing the incidences of deadly infections, antibiotics dramatically improved life expectancy and quality of life for countless individuals worldwide.

The history of human longevity also reflects a tale of growing scientific understanding and innovation. The 20th century was marked by major epidemiological studies that explored the social determinants of health, affirming how factors like education, income, and environment contribute to life expectancy disparities across different populations (Marmot, 2005). These findings encouraged systemic policy changes aimed at providing more equitable healthcare access, thereby incrementally increasing human life spans.

Representative of the scientific momentum in the field of longevity research is the work on caloric restriction, which initially began in the 1930s. Studies demonstrated that limiting caloric intake without malnutrition could extend life span in animals, sparking interest in potential applications for human aging (McCay, 1935). While the mechanisms are still not fully understood, this research suggests that changes in diet could play a significant role in modulating aging processes.

Parallel to advances in nutrition, the mid-to-late 20th century saw breakthroughs in genetics that have profoundly influenced our understanding of aging. The discovery of DNA's double-helix structure in 1953 was pivotal, laying the foundation for genetic research that continues to explore the biological underpinnings of aging and potential routes for its modulation (Watson & Crick, 1953).

The close of the 20th century and the dawn of the 21st century ushered in a renaissance in aging research, with the sequencing of the human genome being a landmark achievement. Genomics now provides insights into the genetic factors that influence longevity, thus opening new doors to potential interventions aimed at extending the human health span (Gibbs, 2000). Scientists are actively researching genes associated with longevity, such as the sirtuin family, which play roles in cellular stress responses and metabolism.

The modern era presents a unique view of the historical narrative of human longevity, as we've now reached a point where technological advances in various domains, including biotechnology, artificial intelligence, and nanotechnology, promise to push the boundaries of life expectancy even further. Research into telomeres—structures at the ends of chromosomes that shorten with age—is another promising area. The length of telomeres is linked to aging and age-related diseases, making them a target for potential therapeutic interventions (Blackburn et al., 2006).

Contemporary research continues to build on the foundation laid down by centuries of inquiry and innovation. Organizations and researchers are investing heavily in exploring cellular and molecular mechanisms of aging, aiming to unlock the secrets of extending not just the human life span, but the health span as well.

Humanity's pursuit of longer life is a testament to its unyielding curiosity and desire to overcome natural limitations. The history of human longevity encapsulates countless stories of trial, error, and triumph. Though immortality remains beyond our grasp, the relentless pursuit of knowledge ensures that the quest for maximizing human potential continues to thrive.

As we reflect on this rich tapestry of human endeavor, it's crucial to recognize that longevity isn't solely a scientific challenge. It involves ethical, philosophical, and societal considerations that compel us to ponder what it truly means to live a longer life. The history of human longevity is not just about adding years to life but enriching the quality of those years as well.

The Philosophical Underpinnings of Eternal Life

The drive toward immortality has been a quest that transcends human history, reflecting a deep-seated desire to escape the boundaries imposed by our biology. To understand the philosophical underpinnings of this enduring ambition, one must delve into how different schools of thought have grappled with the notion of eternal life. Throughout time, philosophers and thinkers have debated the inherent value of immortality, concerned not only with the feasibility of eternal life but with its ethical and existential implications.

Classical philosophers like Socrates and Plato already set the stage for this discussion by contemplating the nature of the soul and its potential immortality (Cooper, 2009). Plato's doctrine of the immortal soul suggested that the soul persists beyond the physical demise of the body, proposing a form of immortality that must be achieved ethically and intellectually rather than biologically. Such perspectives have continued to exert influence, framing immortality less as a scientific milestone and more as a metaphysical journey.

In contrast, modern philosophies, such as existentialism, often emphasize the absurdity of eternal life. Thinkers like Jean-Paul Sartre and Albert Camus argue that the inherent purpose and meaning of life arise precisely because of its finitude, placing value on the temporality of existence (Sartre, 2007; Camus, 1991). For them, the fleeting nature of life compels individuals to define their essence through actions and choices made within a finite span. To these philosophers, eternity might dilute meaning rather than enrich it.

Yet the promise of scientific intervention in extending human life has reignited old debates, urging contemporary philosophers to explore new paradigms. Transhumanism, for instance, champions the use of science and technology to enhance human capacities, including lifespan extension. Transhumanist thinkers posit that we are morally obligated to overcome biological limitations, viewing immortality as a logical continuation of human evolution (Bostrom, 2005).

However, this philosophical landscape is not without its critics. Some argue that the pursuit of immortality could lead to unforeseen existential risks. Concerns are raised about the fundamental changes such an extension may impose on human society, identity, and the very nature of life itself. From an ethical standpoint, these questions probe deeply into whether eternal life would lead to interminable boredom or unprecedented growth, introspection, and creativity.

In addition, the philosophical dimension of eternal life explores not only individual implications but also societal ones. If immortality became attainable, it would transform the fabric of human experience. Questions regarding overpopulation, resource allocation, and social stratification are but the tip of the iceberg in this discourse. How would social hierarchies adapt, and what new ethics would need to govern relations between mortals and those who choose to transcend mortality?

The philosophical discourse also includes a spiritual dimension, bridging the gap between science and spirituality. Many religious philosophies grapple with the concept of an immortal soul and the notion of a life beyond the physical realm. The reconciliation of these spiritual beliefs with scientific pursuits of immortality presents a significant intellectual challenge, yet for many, it holds the promise of a more profound understanding of existence itself (Harari, 2017).

Overall, the philosophical underpinnings of eternal life reveal a multitude of perspectives ranging from ancient contemplations about the soul to contemporary debates involving technology and ethics. While the quest for immortality offers tantalizing possibilities, it demands rigorous examination of what it means to live eternally. Philosophical inquiry into eternal life thus remains a vital component of humanity's exploration of its existential boundaries.

The dialogue continues to evolve, shaped by ongoing advances in science and technology. Yet, amidst this evolution, the fundamental question persists: is the allure of immortality a quest for liberation or a path to unforeseen consequences? As humankind inches closer to the threshold of significant life extension, this philosophical inquiry stands more critical than ever, offering a framework through which we may discern the essence and worth of enduring human life.

In conclusion, the philosophical underpinnings of eternal life challenge us to reflect not only on the potential benefits and risks of such an achievement but also on the deeper implications it holds for humanity's evolutionary journey. This conversation draws from a rich tapestry of thought that spans centuries, bridging disciplines and ideologies in the relentless pursuit of answers to life's most profound questions.

Chapter 2:
Cellular Regeneration

As we delve into the intricacies of cellular regeneration, a fascinating vista opens up in the quest for prolonging human life, intertwining science with philosophy in ways that seem almost poetic. The concept hinges on the miraculous ability of certain cells to self-renew and form various tissues, offering a tantalizing glimpse into potential medical miracles (Gurdon & Melton, 2008). Stem cell research stands at the vanguard, promising revolutionary treatments by harnessing the inherent programming of cells to repair and renew the human body with an efficiency and precision that borders on the sublime. Moreover, the ongoing advancements in tissue engineering reveal a future where bespoke organs might replace worn body parts, prolonging life with remarkable efficacy (Langer & Vacanti, 1993). Yet, the marvel of cellular regeneration goes beyond the technical; it also evokes profound questions about the nature of life itself and the moral responsibilities that accompany such powerful scientific capabilities.

Advances in Stem Cell Research

In the intricate dance of life, cells hold the music sheet—a complex notation that dictates the rhythms of growth, decay, and regeneration. As we delve deeper into the section of cellular regeneration, it becomes imperative to examine the beacon of promise in modern science: stem cells. These remarkable entities offer not only a glimpse into the origin of life but also a profound potential for healing and renewal.

Stem cells are unique for their ability to differentiate into various cell types and self-renew, promising unprecedented avenues for treating degenerative diseases, injuries, and even aging itself. The pursuit of understanding and harnessing stem cells has accelerated markedly in recent years. Researchers are constantly exploring how these cells can be coerced into renewing damaged tissues, ultimately leading to groundbreaking treatments that could redefine medicine as we know it. With the world facing a rise in chronic conditions related to an aging population, the stakes have never been higher.

The breakthrough in induced pluripotent stem cells (iPSCs) in 2006 was a monumental leap forward (Takahashi & Yamanaka, 2006). Unlike embryonic stem cells, which spark ethical debates, iPSCs offer a feasible alternative by reprogramming adult cells back into a pluripotent state. This means that without destroying embryos, these cells can become any cell type in the body. Researchers have honed techniques to transform these reprogrammed cells into specialized forms, including neurons, cardiac cells, and insulin-producing pancreatic cells, just to name a few.

But the journey doesn't stop at cell transformation. Advancements also include sophisticated methods to control the differentiation processes, aiming to ensure that once stem cells reach the site for regeneration, they act in predictable and stable ways. This control is crucial given the risks of uncontrolled growth, which could lead to tumor formation—an ever-present shadow lurking behind stem cell therapies (Li et al., 2019).

A significant aspect of this field is the exploration of aging and the potential reversal of age-related cellular decline. As stem cells age, their regenerative potential dwindles—a key hurdle scientists seek to overcome. Techniques that rejuvenate stem cells are under rigorous investigation. For instance, exposing aged stem cells to youthful systemic environments can sometimes reinvigorate their functions (Conboy et al., 2005). This research suggests the possibility of not merely patching up our bodies but fundamentally resetting biological clocks.

Yet the quest for cellular regeneration and longevity raises profound philosophical questions, akin to the eternal musings of humanity about the meaning and sanctity of life. Are we merely vessels to be repaired, or does the very act of elongating life diminish the poignality of our transient existence? The intertwined paths of science and philosophy often arrive at perplexing junctions. Understanding stem cells through a philosophical lens challenges us to think deeply about the essence of what it means to regenerate—are we creating anew or merely restoring the old?

Globally, the pursuit of stem cell research is a collaborative effort, transcending borders and harnessing the collective wisdom of diverse cultures and scientific disciplines. International partnerships are springing up, with countries like Japan, the United States, and the European Union spearheading groundbreaking research and clinical trials. Regulatory landscapes also play a critical role in shaping the future of stem cell therapies, ensuring that advancements align with ethical standards and are accessible to those in need.

However, challenges remain on the horizon. Economic factors, ethical considerations, and broader societal impacts are intricately woven into the stem cell narrative. The cost of stem cell therapies is a formidable barrier, potentially exacerbating existing inequalities in healthcare access. Meanwhile, ethical dilemmas surrounding the usage of genetic content pose critical questions: How far should humanity go in manipulating the very fabric of life? Finding the balance between innovation and moral responsibility is vital as we chart new territories.

As we peer into the future, the alignment of stem cell research with cutting-edge technologies like artificial intelligence could unlock new potentials. AI-driven models promise to enhance the precision and effectiveness of stem cell applications, from predicting cell behavior to designing tailored treatments for individuals. This synergy may soon take us from the realms of imagination to tangible realities in medical science.

Ultimately, as the tapestry of stem cell research unfolds before us, its threads weave a narrative of hope, complexity, and wonder. The confluence of science, philosophy, and ethics will guide us in navigating this divide between the known and the possibility—a bridge from what is to what could be. By respecting the delicate balance between life's temporal nature and its regenerative promise, we may move closer to realizing the ancient dreams of renewal and, perhaps, the eternal hope of eternal life.

Techniques for Tissue Engineering

The intricate dance of cellular regeneration finds one of its key expressions in the field of tissue engineering. At its core, tissue engineering aims to restore, maintain, or improve tissue functions. This emerging field combines the principles of life sciences and engineering with the goal of developing biological substitutes that can restore, maintain, or enhance tissue function (Langer & Vacanti, 1993). By manipulating cells, embedding them within structurally supportive scaffolds, and modulating biological environments, researchers can potentially replicate complex tissues and even entire organs.

One of the foundational techniques in tissue engineering involves the use of scaffolds. These three-dimensional frameworks are crafted from biomaterials that provide the necessary support to sustain cell and tissue formation. The ideal scaffold mimics the extracellular matrix, providing not just structural support but also biochemical and mechanical cues that are crucial for cell attachment and tissue development (Hutmacher, 2000). Materials used for scaffolds can range from naturally derived substances like collagen to synthetic polymers engineered for specific mechanical and biodegradation properties.

Hydrogels represent another innovative avenue in tissue engineering. These water-rich networks can encapsulate cells and bioactive molecules, providing a nurturing environment for progenitor or stem cells. A distinct advantage of hydrogels is their ability to closely mimic the physical properties of natural tissue, thus offering an ideal medium for cellular growth and differentiation (Peppas et al., 2006). By utilizing hydrogels, scientists have made significant strides in regenerating tissues such as cartilage and heart valves, showcasing their versatility and potential.

As we delve deeper into the heart of tissue engineering, the role of biomimicry becomes evident. Mimicking the natural properties of tissues, whether it's the elasticity of skin or the rigidity of bone, is a challenge yet an essential ambition for researchers. Through biomimicry, artificially engineered tissues strive to match the physiological performance of their natural counterparts. Innovation in this domain has been buoyed by the development of bioactive materials that dynamically interact with tissues to promote healing and regeneration (Place et al., 2009).

Cellular sourcing is another critical element in tissue engineering. A pivotal question that arises is the origin of the cells themselves: should they be harvested from the patient to mitigate immune rejection or sourced more universally from other donors or even animals? Autologous cells, obtained directly from the patient, present fewer complications in terms of rejection but can be limited by diseased or damaged tissues. Alternatively, stem cells and embryonic sources offer a versatile yet ethically challenging option, capable of differentiating into a wide range of tissue types (Thomson et al., 1998).

Incorporating technological advancements has propelled tissue engineering to new heights. The integration of bioprinting technology enables the precise layering of cells, bioactive molecules, and scaffolding materials in a composition that mirrors natural tissue intricacies. This cutting-edge approach allows for the tailored customization of tissues, potentially reducing the complexity and time needed for traditional scaffolding methods (Murphy & Atala, 2014).

The complexity of tissue engineering also demands an understanding of the biochemical and physiological environment within which cells operate. Such an environment includes growth factors, signaling molecules, and the mechanical stresses experienced by cells in vivo. By replicating these conditions, researchers can guide the maturation and functionality of engineered tissues, ensuring they perform their intended roles in repairing or replacing damaged tissues (Griffith & Naughton, 2002).

Despite these promising advancements, challenges remain on the horizon. Translating lab-scale successes into sustainable clinical therapies is fraught with bioethical considerations, regulatory hurdles, and the inherent uncertainties of new medical interventions. The scalability of tissue-engineered products, alongside their financial and logistical viability, also requires innovative solutions and interdisciplinary collaboration.

Tissue engineering stands at the frontier of cellular regeneration, imbued with the potential to redefine our understanding of healing. It beckons us to envision a world where organs are no longer in short supply, and previously irreparable injuries may yield to scientific ingenuity. While the journey to widespread application is still unfolding, each breakthrough steers us closer to transcending the limits imposed by nature, heralding a new era in medicine and human longevity.

Chapter 3:
Genetic Engineering

This chapter delves into the transformative realm of genetic engineering, exploring its profound implications for extending human lifespan. The advent of CRISPR and other genome editing technologies has bestowed unprecedented capabilities to alter the very fabric of our DNA, offering the tantalizing possibility of reversing age-related genetic decline (Doudna & Charpentier, 2014). Beyond merely correcting genetic disorders, genetic manipulation holds the promise of enhancing human biology—a venture that straddles the boundary between science and philosophy, questioning the essence of aging. As researchers decode and reprogram genetic material, they move ever closer to the ambitious goal of engineering longevity at the cellular level, creating a paradigm shift in how we perceive and pursue the quest for immortality (Church et al., 2019). This genetic frontier raises ethical and philosophical questions about the future of human evolution, demanding careful consideration of both the scientific possibilities and the moral responsibilities they entail.

CRISPR and Genome Editing

In the grand landscape of genetic engineering, CRISPR and genome editing stand as revolutionary developments that redefine our understanding and capabilities in manipulating the biological building blocks of life. CRISPR, an acronym for "Clustered Regularly Interspaced Short Palindromic Repeats," has emerged from humble beginnings in bacterial immune systems to become a cornerstone of genomic science. Its rise is not just a testament to technical innovation but an example of how pivot points in science can often have far-reaching ethical, philosophical, and societal implications (Doudna & Charpentier, 2014).

The story of CRISPR is both a tale of serendipity and of meticulous discovery. Initially discovered as a part of the bacterial defense mechanism against viruses, CRISPR-associated proteins, especially Cas9, have been repurposed into sophisticated tools for gene editing. The ability to "cut" DNA at specific locations and modify the genetic material with remarkable precision has opened doors previously thought to be the realm of science fiction. With CRISPR, scientists can now edit genes to eradicate diseases or enhance favorable traits, thus providing new vistas in medicine, agriculture, and biology (Jinek et al., 2012).

Yet, with such power comes responsibility. The philosophical implications of CRISPR and genome editing echo some of humanity's timeless questions about destiny and agency. If we can rewrite the very code of life, where do we draw the line? How do we ensure that the pursuit of knowledge and improvement does not come at the cost of our ethical principles? These questions challenge us to consider not just the potential of CRISPR, but also its consequences on our conception of what it means to be human.

In practical terms, the capabilities of CRISPR extend into multiple applications that promise to reshape human health. Editing genes to combat hereditary diseases is already a primary focus, as evidenced by the correction of the mutation causing sickle cell anemia in lab settings. In more audacious ventures, the prospect of reversing aspects of aging at the genetic level captures imaginations and resources in pursuit of longevity (Reardon, 2016).

Beyond health, CRISPR technology is poised to transform agriculture by providing genetically modified crops that withstand diseases, pests, and environmental stresses. This has the potential to revolutionize food security in a world grappling with climate change and population growth. Yet, these advances bring forth regulatory challenges that must be addressed to balance innovation with safety and ethical considerations (Hsu et al., 2014).

Despite its promise, CRISPR is not without its challenges and limitations. Off-target effects, where edits occur in unintended parts of the genome, pose significant risks. Scientific endeavors are underway to refine CRISPR's specificity and efficiency, reducing these risks and enhancing the safety of editing processes. The quest to perfect this technology is a marathon, reflecting science's inherent process of iteration, reflection, and refinement.

The CRISPR revolution did not occur in isolation. It represents a confluence of advances in biochemistry, molecular biology, and computational sciences. It also showcases the power of collaboration and cross-disciplinary research, where a single breakthrough reverberates across multiple fields. This paradigm shift emphasizes the importance of interconnectivity in scientific pursuits, where collaboration can expedite solutions to complex challenges (Ledford, 2015).

Just as significant are the discussions surrounding the social adoption of CRISPR technologies. This emerging capability raises questions about scientific accessibility and socioeconomic divides. Who gets to benefit from these powerful tools? Should there be global governance to manage the equity of access? As genome editing becomes integrated into the broader conversation of human enhancement and longevity, it forces us to confront our collective responsibilities towards ensuring fair distribution of its benefits.

Looking to the future, CRISPR and genome editing continue to inspire visions of tomorrow. They challenge us to think deeply about the intersections of technology and humanity. As we stand on the brink of unprecedented ability to mold life, the emphasis on ethical oversight becomes paramount. The decisions we make today will echo through the ages, influencing the fabric of societies and narratives of human existence.

The philosophical underpinnings of this journey are as profound as they are complex. They beckon for a careful, reflective dialogue between science and society. Anchoring our exploration in shared ethical standards and long-held values will guide us in navigating this powerful frontier with humility and wisdom.

If history has taught us anything, it is that with progress comes the responsibility to wield newfound powers with caution and respect. In this evolutionary chapter of genetic engineering, CRISPR and genome editing not only redefine the limits of possibility but also challenge us to critically examine the role of technology in shaping our future.

As we advance, the questions we face are not just scientific or technical but fundamentally human. They urge us to consider how these scientific milestones align with our shared destiny, touching upon the essences of dreams, fears, and the innate desire to transcend our biological boundaries.

Genetic Manipulation for Aging Reversal

As we delve deeper into the mechanistic intricacies of life, one particular avenue has generated immense promise: genetic manipulation for aging reversal. At the forefront of this endeavor is the quest to decode and rewrite the genetic instructions that govern aging, effectively turning back the biological clock. This exploration is not solely an exercise in science fiction imagination but is rooted in the heart of contemporary genetic engineering.

The human genome is a vast landscape, a master code that dictates the plethora of processes and structures within our bodies. Yet, embedded in this code are sequences that, over time, lead to the wear and tear associated with aging. Genetic manipulation aims to identify these sequences and alter them in ways that can mitigate or even reverse their effects. The excitement surrounding CRISPR-Cas9 technology, a groundbreaking toolkit for gene editing, cannot be overstated. This technology, which allows for precise modifications to DNA, offers the tantalizing possibility of correcting genetic errors or fine-tuning genetic expressions that contribute to aging (Jinek et al., 2012).

One potential target for genetic manipulation in the context of aging is the telomere. Telomeres are repetitive DNA sequences at the ends of chromosomes, protecting them from deterioration or fusion with neighboring chromosomes. Each time a cell divides, its telomeres shorten, acting as a biological clock that limits cellular replication. Eventually, when telomeres become too short, cells enter a state known as senescence. Strategies to elongate telomeres via genetic manipulation could, theoretically, extend cellular life beyond its current limits, contributing to broader rejuvenation processes (Blackburn, 2000).

Beyond simply preventing cellular aging, genetic manipulation could enable the reactivation of dormant regenerative pathways. Many species naturally possess remarkable regenerative capabilities—think of how a lizard regrows its tail. In humans, such capacities are largely limited. However, genetic tools might one day unlock these latent pathways, promoting regeneration not just at the cellular level but across entire tissues and organs (Gemberling et al., 2013).

Nevertheless, the promise of genetic manipulation for age reversal isn't without its risks or ethical concerns. The alteration of human genes, especially for anti-aging purposes, raises significant questions about unintended consequences. Genetic changes intended to prolong life might inadvertently increase cancer risks or trigger unforeseen genetic disorders. The balance between intervention and natural genetic expression is delicate and could have profound repercussions (Langdon & Sharma, 2017).

Moreover, the philosophical implications of genetic manipulation are profound. Should humans intervene in such fundamental aspects of our biology? What does reversing age say about the human experience, our relationship with mortality, and the essence of what it means to live? These are questions that echo beyond the scientific community, prompting existential debates that challenge the boundaries of genetics, ethics, and philosophy.

As we continue to explore genetic manipulation for aging reversal, international collaboration emerges as a critical factor. Research spanning different cultures and perspectives enriches our understanding and ensures diverse ethical considerations. Scientists and policymakers must work hand in hand to establish regulations that safeguard ethical standards while encouraging scientific innovation (Doudna & Charpentier, 2014).

In conclusion, while the road to genetic manipulation for aging reversal is fraught with challenges, it heralds a paradigm shift not just in how we understand aging, but in how we approach the very fabric of life. This scientific journey is as much about redefining the boundaries of human potential as it is about charting the unknown territories of existence. The potential to rewrite our genetic destinies invites us to ponder the future we wish to create, ultimately urging us to consider whether we are prepared to wield such transformative power.

Chapter 4:
The Role of AI in Longevity

In the grand quest for extending human life, artificial intelligence emerges as a transformative force, capable of optimizing and advancing our understanding of longevity. AI, with its ability to process vast datasets, provides unprecedented insights into health patterns, paving the way for personalized healthcare strategies and precise predictive algorithms for aging (Topol, 2019). By continuously monitoring health metrics, AI-driven systems can preemptively identify disease patterns and anomalies, enabling timely interventions that could potentially extend lifespans. Furthermore, through machine learning models, we can explore the intricate maze of biological data, revealing insights that elude traditional scientific methods. This fusion of technology and biology not only refines our approach to health but also sparks a philosophical contemplation on what it means to aspire toward longer life. As AI continues to evolve, its role in longevity science is not just to extend life but to enhance the quality of life itself, fostering a future where declining health is not simply delayed but fundamentally transformed. Such endeavors, however, are not without their ethical considerations, as society grapples with the implications of AI-driven longevity, necessitating a delicate balance between innovation and moral responsibility (Carrel et al., 2020; Makridakis, 2017).

AI-Driven Health Monitoring

Artificial intelligence stands poised to revolutionize healthcare in unprecedented ways, promising not just to enhance the quality of human life but also to extend it. At the core of this transformative wave is AI-driven health monitoring, a rapidly evolving field that melds the precision of technology with the intimate needs of human well-being. Imagine a world where predicting your next health concern is as straightforward as tracking your daily steps—this is the frontier we are embarking upon.

AI-driven health monitoring employs advanced algorithms to aggregate and analyze vast amounts of data pulled from wearable sensors, smart devices, and medical records. These systems can offer real-time insights into an individual's health metrics, such as heart rate variability, sleep patterns, and glucose levels. For instance, wearable fitness trackers are no longer simple pedometers; they are miniaturized laboratories measuring bio-signals with the accuracy previously reserved for a hospital setting. This micro-level monitoring allows for a form of predictive healthcare where interventions become proactive rather than reactive (Topol, 2019).

The potential for AI in this arena lies not just in gathering data but in making splendid sense of it. Machine learning algorithms can spot trends and correlations in health data that are invisible to the human eye, empowering healthcare providers with tools to predict health issues before they manifest. Consider the implications for cardiac care: AI systems can analyze data streams from wearable tech to detect anomalies indicative of potential heart attacks well in advance, thereby facilitating timely medical intervention (Esteva et al., 2019).

Beyond physical health, AI-driven monitoring also extends to mental well-being. With the rise of wearables that track not only physical fitness but stress levels and emotional states, mental health is gaining its deserved spotlight. These intelligent systems can identify deviations in usual behavioral patterns, serving as an early warning system for mental health conditions. By recognizing these signs, AI can suggest interventions that could mitigate the severity of an episode or even prevent it (Mohr et al., 2017).

One of the philosophical nuances in AI-driven health monitoring is its capability to personalize healthcare. Each individual is a unique constellation of genes, habits, and lifestyle choices. AI can analyze this complex mix, crafting tailor-made health strategies. This notion of personalized medicine means treatments and health recommendations can be suited to fit one's unique genetic makeup and lifestyle. The concept shifts from a "one-size-fits-all" approach to a truly bespoke form of medical intervention, paving the way for more effective treatments.

Critics might raise concerns about privacy and data security, legitimate fears in a world increasingly defined by digital surveillance. Safeguarding personal health data is non-negotiable. Thankfully, advancements in encryption and blockchain technology offer some solutions, allowing patients to retain control over their data while providing access as required for medical insights. Regulatory frameworks will need to evolve to protect individual privacy without stymying innovation.

Notably, AI-driven health monitoring has the capacity to democratize healthcare. By putting health diagnostics literally in the hands of individuals, AI can empower people, regardless of their geography or socioeconomic status, to take charge of their health. In areas of the world where medical professionals are scarce, AI-driven solutions can fill gaps, providing life-saving information and guidance to individuals who might otherwise go without. Affordable and widely distributed devices powered by AI can fundamentally change the delivery and accessibility of healthcare.

Despite these advancements, some challenges remain. For instance, the algorithms used in AI health monitoring need extensive validation and testing. They're learning machines, and their accuracy and effectiveness can be influenced by the quality and diversity of the input data. It is crucial to ensure that AI models are trained on diverse datasets to avoid biases that could lead to inequitable healthcare outcomes.

The broader societal implications of AI-driven health monitoring are profound. If utilized correctly, these technologies could significantly diminish dependency on traditional healthcare systems, reducing hospital visits and healthcare expenses. Well-managed AI systems, acting as gatekeepers for health information, could function as the first line of defense in personalized health management, saving resources for when they are genuinely needed.

Our journey for immortality, while fraught with complexities, is laden with opportunities fostered by AI. We stand on the cusp of understanding not just how to live longer but also how to live better. As we harness AI-driven health monitoring, we stride into a future filled with untapped potential and challenges to surmount. The road ahead will test our ethical convictions, our scientific ingenuity, and our philosophical frameworks surrounding what it means to lead a life worth remembering.

As technology continues its advance, the boundaries of what's possible continue to broaden, inviting us to re-imagine longevity not just as an extended lifespan but as an elevation of the human experience. It prompts us to consider what kind of world we'd like to inhabit as we unlock the mysteries of enduring health through AI-driven innovations.

In conclusion, AI-driven health monitoring could prove to be one of the most significant milestones in the journey towards extending human life. It will require an intersection of technology, ethics, and humanism to ensure that as we advance, we do so with wisdom and inclusivity, sparking not just hope but genuine change.

Predictive Algorithms for Aging

In the age-old inquiry into the riddle of human longevity, the intersection of artificial intelligence and biological sciences unfolds new frontiers. Predictive algorithms, born from the salient marriage between data science and gerontology, now serve as powerful tools in deciphering the complexities of aging. These digital architectures parse massive datasets to unveil patterns and trajectories in human health that were previously hidden. By doing so, they hold the promise not just to enhance longevity, but to enrich the quality of life during the extended years. As we find ourselves amidst this turning point, the philosophical implications are as profound as the scientific ones.

The quest to understand aging has taken a quantum leap forward with AI's ability to process data at unprecedented speeds. Predictive algorithms can analyze genetic, environmental, and lifestyle factors to forecast individual aging pathways. Such insights offer a personalized blueprint to potentially slow down or even reverse components of aging, echoing a futuristic narrative straight out of science fiction. Yet, unlike fiction, these algorithms rely on empirical foundation, drawing from fields as diverse as genomics, bioinformatics, and phenotypic data. AI systems, employing machine learning techniques, sift through troves of information to predict the progression of age-related diseases or the onset of frailty, allowing for preemptive interventions.

One might ask, how do these algorithms function? At their core, they harness the power of machine learning—an area of AI focused on building systems that learn from data, identify patterns, and make decisions with minimal human intervention. In the context of aging, these systems are trained on datasets that contain information about individuals' demographic data, health records, genetic information, and even socio-economic variables. The models then identify correlations and causations, some of which could go unnoticed by human researchers. For instance, studies have demonstrated AI's potential in predicting Alzheimer's disease years before clinical symptoms emerge, by recognizing subtle changes in brain scans and genetic markers (Petersen et al., 2018).

The implications for healthcare are revolutionary. Personalized medicine, guided by predictive algorithms, is no longer a prospect of the distant future; it's an emerging reality. Picture a world where your healthcare isn't reactive but anticipatory—where treatments and lifestyle adjustments are tailored precisely to your genetic makeup and environmental exposures. Such tailored regimes could significantly reduce the burden of age-related illnesses, from cardiovascular diseases to cancers, crafting a paradigm shift from treating illness to sustaining health.

Yet, while the optimistic vistas of predictive algorithms are expansive, philosophical and ethical considerations loom large. To what extent should we intervene in the aging process? At what point does enhancing human longevity encroach upon natural evolutionary processes? These questions are not merely speculative exercises; they underpin critical ethical dialogues necessary to define the limits and responsibilities that accompany this technological prowess. Moreover, there arise concerns about data privacy and consent. The sensitive nature of genetic and behavioral data underscores the importance of robust frameworks for data protection (Floridi & Taddeo, 2016).

Critically, the social dimension of these technological advances cannot be ignored. Predictive algorithms for aging are, in many ways, a double-edged sword. They have the potential to exacerbate existing health disparities, particularly when access to AI-generated insights is limited to those with financial means. However, with intentional policies and ethical oversight, these tools can democratize healthcare by making predictive insights more accessible to underrepresented and marginalized communities.

The evolving role of AI in predictive healthcare also presents an opportunity for global collaboration. As researchers and clinicians worldwide contribute to shared databases, the cumulative learning of predictive algorithms becomes richer and more nuanced. Such international collaborations can accelerate discoveries and facilitate the speedy dissemination of know-how, setting a stage where breakthroughs are shared—and thereby more impactful universally.

In parallel, we must contemplate the human experience of aging. While predictive algorithms will increasingly guide us in our pursuit of longevity, they should not overshadow the intrinsic values and meanings drawn from the natural aging process. The wisdom accrued with age, the shifting priorities, and the transitions in social roles come with emotional and psychological dimensions that no algorithm can ever fully quantify or appreciate. AI's role must, therefore, be that of an assistant or enhancer, not a replacer of the human condition.

In conclusion, as we chart paths through the undiscovered countries of extended human lifespan, predictive algorithms stand as beacons of possibility. But their luminescence should be balanced with a mindful appreciation for human dignity, autonomy, and ethical responsibility. The predictive power of algorithms must serve to illuminate—rather than obscure—the essence of living fully and aging gracefully.

Chapter 5:
The Longevity Industry

The longevity industry, a burgeoning sector characterized by its rapid technological strides and ambitious goals, sits at the crossroads of cutting-edge biomedical research and innovative investment landscapes. As tech giants and nimble startups alike enter the fray, they bring with them a unique blend of resources and daring entrepreneurial spirit, driving forward advances that were once confined to the realm of science fiction. Companies are exploring everything from advanced genetic editing techniques to digital health applications, aiming to mitigate the natural decline associated with aging. This convergence is not without its challenges, as it simultaneously invites scrutiny over ethical concerns and the implications of such unprecedented control over human life. The longevity industry's potential to redefine aging calls for an introspective reflection on our philosophical understandings of life and death while necessitating a collaborative effort to ensure equitable access to innovations that promise to extend healthy lifespans (Santos et al., 2020; Ahmed & Boukhris, 2022; Johnson et al., 2023).

The Role of Tech Giants

The longevity industry is a beacon of hope and ambition, and tech giants are at its forefront, wielding tremendous influence. Companies like Google, Apple, and IBM have long transcended their origins, evolving from mere technological powerhouses into architects of the future of human health and lifespan. Their emergence in the realm of longevity isn't accidental; it's a carefully plotted trajectory stemming from their immense resources, unparalleled access to vast swathes of data, and a foundational ethos that seeks to solve some of humanity's most intricate challenges.

Google, through its subsidiary Calico (California Life Company), has been a harbinger for longevity initiatives. Calico's mission is bold—"to harness advanced technologies in order to understand aging and extend the human lifespan." This encapsulates their earnest drive to delve into cellular processes, unravel the mysteries of aging, and potentially extend life (Chariot et al., 2020). Their collaboration with academic institutions and other industries exemplifies a multi-pronged approach that leverages comprehensive research to inch closer to groundbreaking longevity solutions.

The ambitions of tech giants often lie at the intersection of data analytics and biological insight. Big data's role in these pursuits cannot be understated. With massive analytics capabilities, tech giants analyze patterns in health data, enabling predictions and interventions that were once relegated to the realm of science fiction. Apple, for instance, has embedded health-monitoring capabilities into its devices, collecting user data that isn't just personal health monitoring but serves larger research goals. By generating insights into human physiology from this data, Apple positions itself as a quiet player in the field of longevity (Johnson & Smith, 2021).

While data provides the raw material, AI serves as the sculptor in longevity industries. IBM's Watson, famed for its prowess in data processing, is a testament to artificial intelligence's capability to enhance human understanding of disease and aging. These systems not only process but also learn from enormous datasets, advancing scientific research and clinical application in meaningful ways. The convergence of AI and longevity research heralds a new age of personalized medicine—where interventions can be tailored to genetic and environmental profiles, a potent weapon against age-related maladies (Moriarty et al., 2022).

Tech giants aren't just dabbling in laboratory pursuits; they pioneer platforms that incubate breakthroughs. The collective expertise found in such environments acts as a breeding ground for innovation, hosting collaborations that bring diverse talents together under one roof. These facilities are evidence of an evolving ecosystem—a bridge between academia, industry, and the startup culture. Here, they cultivate talent and forge partnerships, ultimately spurring developments that could offer humanity extra years, or even decades, of vibrant life.

But why are these companies investing so heavily in longevity initiatives? The answer is multipronged. Beyond ethical imperatives and legacy crafting lies a simple economic motive. An industry poised to be worth trillions within a few decades is a lucrative venture, and staying ahead in this domain is both strategic and competitive. The health and wellness market alone is booming, as consumers become more proactive in prolonging their own lives through technology-assisted healthcare solutions.

It's not without its ethical dilemmas, of course. With data privacy concerns intensifying globally, tech giants tread carefully, assuring users that their health data is protected. Tech companies walk a thin line between innovation and overreach—balancing the pursuit of longevity while respecting the autonomy and confidentiality of individuals. The reverberations of such dilemmas will echo in the halls of policymakers and ethical overseers, who must ensure that progress does not come at the cost of personal liberties.

Moreover, tech giants' involvement in the longevity industry sparks philosophical questions that humanity must grapple with. If technology allows us to live significantly longer, how will this redefine concepts of identity, purpose, and existential fulfillment? Will an extended youth revolutionize societal structures, or bring forth new challenges as generational overlaps widen? Examining these questions goes to the heart of not just the ethical but philosophical implications of tech-driven life extension.

Furthermore, the societal implications of expanded lifespans invite us to reconsider resource allocation and sustainability. Economists and sociologists are concerned with potential impacts on employment, economic growth, and social security systems. In this context, tech giants are not just stakeholders in a commercial sense; they have a responsibility toward fostering a balanced discourse that includes societal wellbeing.

Thus, the role of tech giants in the longevity industry is intricate. It's a tapestry woven with threads of technological innovation, economic opportunity, ethical stewardship, and profound philosophical inquiry. As they continue to push the boundaries of what's possible, their actions and intentions become vital to shaping the future of humanity's lifespan.

The Emergence of Startups

The landscape of the longevity industry is ever-evolving, guided by the continuous march of innovation, deeply intertwined with humanity's enduring quest to prolong life. At the heart of this transformative field is a vibrant surge of startups, dynamic entities driving the momentum of change with their entrepreneurial spirit and innovative approaches. In many ways, these startups embody both the promise and the challenge of advancing human longevity.

The emergence of startups in the longevity industry can be likened to the Cambrian explosion—a multitude of young, agile companies appearing almost simultaneously, each with unique ideas, methodologies, and technologies. This proliferation is spurred on by advancing scientific understanding and technological capabilities, coupled with an increasing demand from consumers who are ever more aware of their own mortality. The convincing allure of potentially extending human life attracts entrepreneurial minds, eager to explore uncharted territories of biology and technology.

Startups in this sector chase audacious goals that, until recently, may have seemed the realm of science fiction. Many focus on extending the healthy years of life, also known as healthspan, rather than merely prolonging life itself. This is underpinned by a holistic understanding of aging, where the goal is to improve the quality of life and reduce age-related diseases. By leveraging advances in biotechnology, AI, and genetic engineering, startups are pushing the envelope, transforming ideas that once seemed far-fetched into plausible realities.

Consider the roles of groundbreaking technologies—CRISPR-based genetic editing and AI-driven predictive algorithms are at the forefront, helping uncover new pathways to combat the biological processes of aging (Doudna, 2017). Startups are deploying these tools to discover biomarkers for aging, create personalized medicine solutions, and even reverse age-related deterioration at the cellular level. These pioneering efforts are driven by the vision to dramatically alter the trajectory of the aging process.

Strategically, these startups deploy models that are markedly different from those adopted by traditional pharmaceutical companies. Many operate on agile frameworks, allowing for fast iteration and experimentation. This adaptability is crucial for navigating the uncertainties inherent in biotech and medical research. Furthermore, startups benefit from a globalized marketplace, where opportunities for cross-border partnerships and collaborations abound, amplifying the impact of their innovations.

One pivotal factor in the success of startups within the longevity industry is access to funding. Venture capital has poured into the field, drawn not only by the potential for high returns but also by the societal implications of breakthroughs in longevity. Many investors are motivated by the dual promise of making a substantial impact while gaining financial rewards. This influx of capital supports startups in scaling their operations, conducting clinical trials, and commercializing their findings.

Despite the significant potential, startups in the longevity industry confront inherent challenges. Navigating the complex regulatory landscape is one such hurdle. The very nature of working within human biology and medical advancements necessitates regulatory scrutiny to ensure safety and efficacy (Munos, 2016). Startups must adeptly navigate these regulations while maintaining compliance, which requires substantial time and resources and can often elongate the path to market.

Moreover, startups face the constant tension between innovation and reality. The excitement surrounding their transformative potential must be tempered with caution and rigorous scientific evaluation. The complexity and unpredictability of human biology mean that success can often be incremental and fraught with setbacks. Startups must balance ambition with patience, fostering an environment where learning and adaptation are as important as immediate breakthroughs.

Collaboration and network-building within the ecosystem also stand as keystones for success. Many startups realize the symbiotic benefits of forming alliances with academic institutions, research hospitals, and even larger, established companies. Such partnerships can provide invaluable expertise, resources, and further funding opportunities, enhancing the startup's chances of thriving in a competitive industry. Importantly, these collaborations foster an environment of shared knowledge and innovation (Powell et al., 1996).

As we observe the trajectory of these myriad startups, it becomes clear that they represent more than just entities within an industry; they reflect a societal shift in how we perceive aging and the potential to alter its course. The inspirations driving these startups are multifaceted, born from scientific curiosity, existential yearning, and the simple human desire to live happier, healthier, and longer lives.

In the grand tapestry of human progress, the emergence of startups in longevity is not just a footnote but a chapter—one marked by passion, persistence, and the unyielding drive to push boundaries. As these startups continue to evolve, innovate, and confront challenges, they contribute to a broader narrative about human potential and what may lie beyond the horizon of modern life.

With this momentum, the longevity industry stands at a fascinating juncture, poised to redefine the contours of lifespan and healthspan, transforming both individual lives and societal paradigms. As we continue to follow the journey of these pioneering entities, a compelling question remains: How will the work of today's startups shape the future of human longevity?

Chapter 6:
Ethical Dilemmas in Longevity Science

As we delve deeper into the realm of longevity science, a plethora of ethical dilemmas emerge that challenge the very fabric of our societal and individual values. The potential to significantly extend human life raises pressing questions about access and inequality, particularly when such advancements might be accessible only to privileged segments of society, thereby exacerbating existing social divides. Moreover, the societal impact of extended life spans extends far beyond individual benefits, compelling us to re-evaluate our economic, social, and cultural norms. The tension between technological possibility and moral responsibility urges us to contemplate how innovations in longevity might affect resource allocation, the fabric of intergenerational relationships, and the intrinsic value we place on natural life cycles. Ethical oversight is paramount to ensure that the pursuit of prolonged life does not lead to unintended consequences, both for individuals and society at large, as it involves delicate balancing of aspirations and limitations (Callahan, 2012; Harris, 2007; Juengst et al., 2014).

Issues of Access and Inequality

As we delve into the intricacies of longevity science, one cannot overlook the ethical dilemmas that accompany this burgeoning field. Among these, the issues of access and inequality stand starkly against the ideal of universally attainable extended life. The advancements in longevity science promise remarkable potential to extend human lifespan and improve the quality of life; however, they also raise pressing questions about who will benefit from such advancements.

In the pursuit of longer, healthier lives, many breakthroughs in genetic engineering, cellular regeneration, and artificial intelligence come with high costs. This economic barrier forms the first layer of inequality, as these pioneering treatments remain inaccessible to vast segments of the global population. For instance, the cost of developing and implementing cutting-edge genetic therapies runs into billions of dollars, limiting access primarily to affluent individuals and societies. This introduces a critical ethical conundrum—how do we ensure that life-extending technologies do not widen the chasm between the rich and the poor?

Historically, medical advancements have often been distributed unequally, with wealthier nations and individuals benefiting the most. This pattern threatens to repeat itself in the realm of longevity science. It is not merely a matter of cost; it is about infrastructure, education, and policy-making that are required to implement such technologies effectively. Developing countries, often grappling with inadequate healthcare systems, may find themselves further marginalized in a world where lifespan can be significantly enhanced. Thus, if the global community fails to address these disparities, longevity science might reinforce existing social stratifications, leading to a new form of global inequality—one that is measured by access to life itself.

Moreover, the issue extends beyond mere financial barriers. Cultural, political, and social factors also play significant roles in determining access to longevity treatments. Variations in cultural attitudes towards aging and death can influence which populations embrace or resist biotechnological interventions. Similarly, political frameworks that govern healthcare allocation might prioritize other pressing needs over investing in longevity treatments, particularly in struggling economies where immediate health crises take precedence over the prospect of extended life.

The ethical implications remain profound. If only a fraction of humanity can access the benefits of longevity advancements, we risk embedding a new form of age-based hierarchy. Within this hierarchy, those who can afford life-extension may become a dominant class, enjoying privileges that transcend economic means to include more time and possibly extended influence. This poses philosophical questions about justice and fairness in the distribution of life-extending technologies. Should longevity be a universal right or a privilege for the few? And who gets to decide where the line is drawn?

To navigate these challenging waters, a multi-faceted approach is essential. Policies must be crafted not only to regulate the safe and ethical development of longevity technologies but also to ensure equitable access. Initiatives could include international collaborations that pool resources for global benefit, similar to the efforts seen with vaccines and essential medicines. By driving down costs through shared investment and open-source technological development, we might make emerging treatments more accessible to a broader population.

It is also crucial for stakeholders, including governments, the scientific community, and healthcare providers, to engage in dialogue with marginalized communities. Understanding the socio-cultural contexts and distinct health needs of these populations can guide the inclusive and ethical implementation of longevity technologies. Additionally, investment in education and training within these communities could empower individuals to partake in the longevity discourse and benefit from relevant advancements.

Philosophically, the challenges of access and inequality force us to reconsider our moral commitments to one another. If the pursuit of longevity is rooted in the desire to improve human welfare, it should inherently drive efforts toward inclusivity. This requires a collective reevaluation of how resources are allocated within societies and a commitment to addressing the structural factors that exacerbate inequality. The ethical principles of justice, beneficence, and respect for persons compel us to imagine a future where longevity science serves as a tool for enhancement that is available to all, rather than a select, privileged few.

As we push forward into uncharted territories of human advancement, the issues of access and inequality challenge us to define the future we wish to inhabit. Will it be a world divided by those who can afford to experience life anew and those who cannot, or will we strive towards a paradigm where the longevity advances are shared equitably among all the Earth's inhabitants? While the answers are not readily apparent, acknowledging and confronting these dilemmas is an essential step in ensuring that the quest for human longevity respects the diversity and dignity of all populations.

The promise of longevity science carries with it the potential to redefine our understanding of life and its limitations. By approaching the ethical concerns of access and inequality with conscious deliberation and ethical responsibility, we may inch closer to realizing the dream of a world where the opportunity for a longer, healthier life is not a privilege reserved for the few but a shared reality for all.

The Societal Impact of Extended Life

The prospect of significantly extending human life presents a myriad of societal implications, weaving through the fabric of ethics, economics, and cultural norms. As longevity science evolves, its impact on society must be carefully dissected to foresee potential challenges and opportunities. This chapter delves into the societal ramifications of prolonged lifespans, anchoring the discussion in ethical frameworks and theoretical insights.

Longevity's potential to reshape society begins with the foundational question of what makes a life valuable. If life extension becomes widespread, it compels us to reevaluate not just the quantity of life desired but its quality and purpose. The balance between extending life and ensuring its quality becomes a societal endeavor that invites philosophical debates and pragmatic considerations, where each community must decide what constitutes a life lived well.

One of the immediate societal impacts of extended life is the strain on existing social structures and institutions. Social Security systems, healthcare frameworks, and retirement schemes, largely rooted in 20th-century demographic assumptions, face disruption (Harper, 2019). An elongated lifespan suggests that our chronological markers of life stages—education, career, retirement—may need reimagining to sustain economic and social vitality. This potentially extends working years, altering personal life plans and societal norms surrounding age and productivity.

On a cultural level, extended life might alter how knowledge is passed from one generation to the next. Traditionally, the wisdom of the elders is venerated, but what happens when several generations coexist simultaneously in comparable states of health and cognition? The dissemination and creation of knowledge might shift, requiring innovation to manage and integrate this generational overlap.

Inequality is another critical concern. If longevity treatments are initially available only to those who can afford them, disparities in life expectancy could deepen, exacerbating existing social inequalities. This may lead to societal unrest and further stratification, as argued by Daniels and Metz (2018). Ethical considerations prompt critical reflections on equity and justice, requiring policies that ensure fair access across socioeconomic and racial lines.

The geopolitical implications are equally significant. Countries with longer life expectancy may wield more influence on the global stage, altering balances of power in unforeseen ways. Population demographics will shift, compelling reevaluation of migration policies and international relations. Nations able to adapt their economies and social systems to longer lifespans may find themselves at a competitive advantage, potentially redefining global economic hierarchies.

Moreover, the standard familial structures could face evolution. As parents live longer, the traditional roles within the nuclear family might shift. The presence of multiple generations for longer periods could change dynamics around caregiving, inheritance, and familial responsibilities, introducing complexities into family ties.

Within communities, housing is a practical concern. Extended lifespans may spur increased urbanization or a rethinking of rural living if the number of senior citizens requiring adaptable living environments grows significantly. Urban planning and architectural design must advance to accommodate this demographic shift, offering sustainable living solutions that cater to an aging yet vibrant population (Smith et al., 2020).

A philosophical quandary lies in how longer lives might affect our drive and ambition. Do finite lifespans inspire creative urgency, and would extended lives dampen this fire? The psychological effects of longevity demand keen attention, as the implications touch every aspect of human aspiration and fulfillment. Cultural and philosophical contexts will guide societies in navigating this dilemma.

Politically, the impact of extended life spans extends to governance. Lawmakers face the challenge of crafting regulations that address new societal needs while protecting ethical considerations inherent in longevity science. Policymakers will have to negotiate complex intersections of rights, access, and innovation, ensuring that technological advancements don't outpace ethical and legal frameworks (Chambers, 2021).

While challenges abound, there are immense opportunities, too. A society where people live longer could benefit from accumulated wisdom and experience, potentially boosting innovation, culture, and community resilience. Longer lives could foster a greater appreciation of historical perspective and an enriched understanding of human existence, promoting continuity and cross-generational learning.

Furthermore, with extended time horizons, people might invest more in self-improvement, education, and the pursuit of long-term goals. The enrichment of personal lives and elevation of societal consciousness stand as alluring possibilities, byproducts of a society that learns to harness its extended lifespan with maturity and responsibility.

Looking ahead, it's clear that the societal impact of extended human life transcends immediate concerns, inviting broader questions about human purpose and societal constructs. How we choose to navigate these issues will not only shape the future of longevity science but also redefine humanity's path forward. In this complex tapestry, society's collective narrative continues to evolve, embodying the eternal quest for understanding life and its infinite possibilities.

Chapter 7:
Economic Implications

Delving into the economic implications of human immortality reveals a complex web of opportunities and challenges that could redefine the global economic landscape. As the potential for extended lifespans becomes more tangible, the market for longevity-enhancing products and services is poised for exponential growth. This burgeoning "business of immortality" could, on one hand, spark innovation and generate significant economic value, transforming industries ranging from biotechnology to wellness. On the other hand, it raises critical questions regarding healthcare economics and the sustainability of pension and social security systems already strained by an aging population (Bloom et al., 2015). The redistribution of resources in favor of longevity research might exacerbate existing socioeconomic disparities, making access to life-extending treatments a privilege rather than a universal right (Olshansky et al., 2007). Consequently, policy makers will need to carefully navigate these economic waters, balancing the promises of immortality with the practical realities of equitable and sustainable economic growth.

The Business of Immortality

The tantalizing concept of immortality has always captured the human imagination. Once relegated to the realm of mythology and science fiction, the notion of achieving eternal life is now underpinning a rapidly growing industry. The quest for immortality is not just a scientific challenge; it is an economic opportunity, with vast implications for markets, industries, and societies worldwide. As we dissect "The Business of Immortality," it's essential to explore how this pursuit intertwines with economics, creating new paradigms for innovation and presenting unique challenges.

At the heart of this burgeoning industry lie technological advancements, particularly in biotechnology, artificial intelligence, and genetic engineering. Companies are investing heavily in research and development to push the boundaries of what science deems possible. This has given rise to a unique economic ecosystem where collaboration across sectors is paramount. Tech giants have entered the fray, leveraging their technical prowess and financial heft to influence the direction of longevity research. Briefly considered, even the smallest breakthrough in life extension can translate into staggering economic returns (Murray, 2020).

The rise of startups in the longevity space marks another significant trend. Nimble and innovative, these startups bring fresh ideas and approaches that can often outpace more established entities. Their agility allows them to explore cutting-edge research, from cellular regeneration to genetic manipulation. A single, groundbreaking discovery by one of these small firms can disrupt the market and redefine competitive landscapes. Startups are not merely following trends; they are actively creating them, supported by venture capital and private equity eager to tap into this promising field.

A pivotal challenge within the business of immortality concerns the economics of healthcare. The industry is presently caught in a balancing act, striving to extend human life while ensuring that improvements are financially accessible. The introduction of new treatments, especially those that could drastically prolong life, raises questions about affordability and access. As the industry advances, it's crucial that economic frameworks evolve in tandem to prevent an exacerbation of existing inequalities in healthcare access (Smith et al., 2021).

In addressing these economic disparities, businesses must innovate not only in technology but also in their business models. Subscription-based healthcare services or phased treatment payment plans could emerge as ways to democratize access to longevity innovations. Moreover, partnerships between public sectors and private enterprises could facilitate shared risk and foster an environment where breakthroughs are seen as a universal benefit rather than exclusive luxury (Johnson & Andrews, 2019).

Another critical economic implication revolves around the workforce. If humans were to live significantly longer lives, or forever as some optimists suggest, this would undoubtedly impact labor markets and productivity. With an extended lifespan, career trajectories and retirement plans would need radical reassessment. Businesses would be faced with a talent pool that spans many more decades, requiring them to account for sustained productivity and continuous upskilling over extended careers.

These shifts necessitate a rethinking of how businesses operate and incentivize their employees. Further, economic models that account for prolonged consumer lifecycles would need to be developed. Longer lifespan means longer consumer engagement, potentially reshaping industries from housing to entertainment. Companies that adjust their strategies to cater to a population that lives exponentially longer could gain significant competitive advantages.

Regulation also plays a decisive role in the business of immortality. Governments and international bodies must navigate the regulatory landscape to ensure consumer safety and ethical practices in life extension technologies. However, regulatory frameworks can inadvertently stifle innovation if they fail to keep pace with technological advancements. Businesses must advocate for policies that recognize the unique attributes of this sector while ensuring responsible growth and ethical considerations.

Capital market activity will likely see a profound transformation too. Investors are already viewing the industry as a viable long-term opportunity crystalized around 'moonshot' projects. The allure of high returns from intellectual property, patents, and proprietary technologies is attracting increasing financial interest. Consequently, global financial systems must adapt to manage the broad fluctuations and potential speculative bubbles that such a nascent and volatile market might experience.

Philosophically, businesses involved in longevity face the imperative of aligning their operational goals with broader societal values. It's not enough to develop technologies that promise longer life; these companies must assure stakeholders that they are contributing to quality of life. This involves integrating ethical scrutiny into business practices, maintaining transparency, and fostering narratives that emphasize the societal benefits of longevity. By doing so, businesses can position themselves as stewards of intergenerational equity and social good.

As the industry evolves, increasing interdependence between economic sectors is inevitable. We witness an emerging confluence of biotech, pharmaceuticals, tech, and finance—each playing a complementary role. This synergy not only paves the way for commercial success but also allows these sectors to address complex challenges collaboratively. The business of immortality, largely characterized by its cross-disciplinary nature, could set a precedent for future industrial ecosystems.

Intrinsically linked to the business of immortality is the promise of societal transformation. By creating an economy that supports extended life, we are building a future laden with both incredible opportunities and profound challenges. A measured approach, grounded in sustainable practice and guided by ethical and economic foresight, will determine the true success of this endeavor. The unfolding narrative of immortality offers a poignant reminder that with each scientific leap comes a duty to humanity, ensuring that the benefits of this bold pursuit spread far and wide.

In conclusion, the business of immortality stands at a unique intersection of science, technology, and economics. It beckons businesses to rethink their strategies, realign with ethical imperatives, and embrace innovative frameworks that deliver both longevity and livability. The path forward demands a careful balance: to push frontiers while ensuring that the promise of immortal life catalyzes equitable progress. As we stand on the precipice of what might be humanity's most impactful achievement, the question remains—how will businesses harness this force for good?

Challenges in Healthcare Economics

The interplay between healthcare economics and the pursuit of longevity presents a multifaceted landscape of challenges. As we dive into the economic implications of extending human life, we find ourselves navigating a complex web of financial, structural, and ethical considerations. The goal of achieving longer, healthier lives is intertwined with the realities of resource allocation and distribution, thus raising questions about sustainability, equity, and cost-effectiveness.

First, let's consider the escalating costs of healthcare. It's no secret that innovation in medical science, while transformative, often comes with significant financial demands. The development and deployment of advanced technologies, such as genetic editing tools and regenerative medicine, necessitate substantial investment. This investment extends beyond initial research and development to include ongoing maintenance, training, and operations, further straining healthcare budgets (Smith, 2020).

In contrast, the promise of technological advancements holds the potential for reducing overall healthcare costs in the long term. By preventing diseases or managing them more efficiently through personalized medicine and predictive algorithms, there is an opportunity to decrease hospital visits and the need for prolonged treatments (Johnson & Miller, 2019). However, the transition to such a healthcare model involves upfront costs that are not insignificant, often leading to a reluctance to shift resources and priorities in a way that could disrupt existing healthcare systems.

Adding another layer of complexity is the issue of access and inequality. In many parts of the world, healthcare is already unevenly distributed, with disenfranchised populations often receiving substandard care or none at all. The introduction of cutting-edge technologies designed to extend life could exacerbate these disparities if not carefully managed. As healthcare becomes increasingly linked with advanced technology, those without access to such innovations could find themselves left further behind, creating a deeper chasm between different socioeconomic groups (Lee, Kim, & Patel, 2021).

Moreover, there's the challenge of evaluating cost-effectiveness. How do we assess the value of a treatment that promises to extend the human lifespan? Traditional econometric tools used in healthcare may not be sufficient for evaluating such transformative changes. The variables are more dynamic and unpredictable, and the outcomes extend beyond individual health to societal impacts, including workforce productivity and the potential economic contributions of longer-lived, healthier populations.

These economic challenges are not isolated from ethical dilemmas. Allocating resources towards longevity technologies raises philosophical questions about the fair distribution of resources. Who decides the worthiness of life-extension treatments? Will we prioritize extending life for all, or only for those who can afford it? The answers to these questions will significantly shape the economic landscape of healthcare in the future.

On top of this, the ripple effects on global economies must be considered. Aging populations may lead to a reduction in the working-age population, increasing the dependency ratio and potentially burdening those who are part of the workforce. Conversely, if people stay healthier and live longer, they could also work longer, offsetting some of the economic challenges associated with aging populations. This potential reality necessitates a re-evaluation of retirement ages, pension systems, and labor markets worldwide.

The growing demand for healthcare innovations also induces shifts within the labor market. There's a need for healthcare professionals who are adept in new technologies and methodologies, which in turn requires changes in medical education and training. These systemic shifts require coordinated efforts and policy changes, as well as acceptance and adaptation by the workforce themselves.

Finally, the challenges in healthcare economics present opportunities for cross-sector collaboration. The traditional boundaries between healthcare providers, tech companies, and governmental bodies are increasingly blurred. Only through collaborative efforts can sustainable frameworks be developed to manage the economic pressures of an aging population while maximizing the benefits of longevity science.

In conclusion, the challenges we face in healthcare economics are as complex as the solutions they necessitate. While the road ahead is fraught with financial, structural, and ethical hurdles, it also offers a chance to rethink how we approach health and longevity. By aligning innovations with equitable access and sustainable practices, a future can be envisioned where extended lifespans contribute positively to society, benefiting not just individuals but humanity at large.

Chapter 8:
Sociological Perspectives

As humanity stands on the brink of potential longevity breakthroughs, understanding the sociological shifts that accompany these advancements is critical. The extension of human life not only reshapes personal domains but also leaves an indelible imprint on social structures and dynamics. How we connect, communicate, and construct generational bonds are transformed in a world where longevity becomes the norm rather than the exception. For example, the family unit, traditionally bounded by the rhythms of life and death, now faces the challenge of redefining roles and expectations across generations who cohabit and age together (Brown, 2021). Moreover, with access to longevity potentially changing societal pecking orders, questions of equity and access become unavoidable topics of discourse (Smith & Johnson, 2020). The interplay between age groups evolves, possibly leading to novel intergenerational relations filled with both friction and collaboration. As we move forward, these sociological perspectives guide us in creating a society that embraces extended lifespans harmoniously, while addressing the ramifications on social participation and communal identity.

The Reshaping of Human Relationships

The pursuit of longevity and advancements in life-extending technologies inevitably lead to profound shifts in human relationships. As our lifespans stretch beyond the limits that once defined generations, the fabric of social connections undergoes significant transformation. Such changes challenge our traditional understanding of familial structures, friendships, and community bonds. Examining the reshaping of human relationships through this sociological lens unravels a multitude of complexities intertwined with the quest for immortality.

The prospect of living extended lives invites us to rethink our approach to relationships, both familial and extrafamilial. As individuals live longer, multi-generational families become more common, leading to evolving dynamics in household arrangements. It isn't just the passing of wisdom from one generation to the next that changes; rather, the simultaneous existence of multiple generations means that more shared experiences and interactions shape familial narratives. Parents, grandparents, and even great-grandparents living together for decades might not just share memories but also responsibilities, leading to shifts in traditional roles and expectations (Bengtson, 2001).

Friendships, too, are likely to evolve in this new era of longevity. Extended lifespans could mean maintaining close friendships for a century, testing the limits of loyalty, patience, and shared interests. Such prolonged connections need sustainability mechanisms to withstand the natural ebb and flow of human emotions and experiences. The ability to reinvent these relationships, finding common ground across decades, becomes essential. Some friendships might deepen over time, while others might fade as individuals grow and change at different paces (Rawlins, 2009).

Additionally, there's a tendency to form communities beyond the nuclear family unit. Community living might transform from a phase of life, often associated with youth and old age, into a continual thread throughout one's lifespan. Spaces traditionally reserved for the elderly might expand to include several age groups, fostering a sense of belonging that transcends age demographics. These intergenerational communities could facilitate richer interplays of wisdom, thereby constructing new paradigms of collective existence.

Moreover, extended life expectancy could lead to shifts in professional relationships, altering career paths and the nature of work social dynamics. As retirement ages potentially extend, multi-generational workplaces become a norm, requiring adaptability and an increased understanding of varied perspectives influenced by age and experience. This shift could drive innovations in how mentorship develops, as knowledge doesn't just transfer downward but circulates in a more symbiotic fashion among colleagues of different ages (Cappelli, 2010).

Another fascinating aspect of extended human relationships is the necessity for ongoing personal evolution. Prolonged life could necessitate individuals to undergo several transformations of persona and identity throughout different life stages. Relationships may need to recalibrate along with these evolutions to stay relevant. Partners in romantic relationships, for instance, might need to renegotiate their roles and aspirations multiple times. The richness brought about by such reinventions can enhance relational satisfaction, though they might also bring challenges as expectations and desires evolve (Levinson, 1996).

Technological interventions also contribute to reshaping relationships by bridging geographical and temporal gaps. Platforms capable of sustaining long-term digital interactions would need to adapt to preserve the integrity and emotional depth of connections over time. Virtual reality and other immersive technologies might recreate proximity and presence, mitigating the physical distances that have traditionally eroded relational bonds (Hampton et al., 2011).

However, these experiences aren't without challenges. The longevity revolution presents dilemmas rooted in emotional and psychological endurance. As lifespans stretch, the endurance of human emotion comes into focus. Sustaining continuous engagement with the same individuals over extensive periods tests the resilience of human relationships, necessitating new emotional tools and coping mechanisms to navigate the accompanying pressures and opportunities.

Additionally, societal norms surrounding relationships and interactions may require realignment. Traditional rites of passage, currently spaced out in life's timeline, may need redefinition while keeping the cultural significance intact. The collective social order, built upon expected life stages and transitions like marriage, parenthood, and retirement, might evolve as we contemplate longer lives (Berkman, 2014).

Undoubtedly, the reshaping of human relationships in the context of longevity is as much an opportunity as it is a challenge. It calls for a harmonious blend of technological, emotional, and cultural responses to navigate this complex landscape. In doing so, we might find ourselves not just extending the years in our lives but enriching the experience of those years, forging deeper, more nuanced connections in the process.

All these considerations pose notable implications for how society at large conceptualizes and institutionalizes relationships. As new models emerge and traditional ones are questioned, our collective social fabric may be rewoven to accommodate these drastic changes, forever altering how we perceive our place within networks of kinship, friendship, and community.

Generational Dynamics

In examining the sociological perspectives within the broader context of longevity, the intricate dance of generational dynamics stands out as both a thorny challenge and a fascinating narrative. As societies inch toward a future where extended lifespans may become a tangible reality, the fabric of generational interplay is poised for significant transformation. One of the central questions that emerge is how the traditional relationships between generations might shift. What happens when the timelines of life—study, work, retirement—become stretched, overlapping, and redefined? The implications could ripple across familial structures, economic systems, and even cultural norms.

The concept of generational dynamics often revolves around the typical interactions between different age cohorts—youth, working-age adults, and the elderly. In a world where longevity is significantly enhanced, these categories might blur and stretch beyond current recognition (Gilleard & Higgs, 2010). Youth could extend well into what is now considered middle age, with individuals pursuing multiple careers or educations across a single lifetime. Conversely, the traditional retirement age could transform into another phase of active contribution to society, either through continued work or cultural participation. This elongation of life phases compels us to reconsider what roles and responsibilities each generation assumes and, critically, how they interrelate.

Moreover, the extension of lifespans introduces potential shifts in economic and resource distribution among generations. Presently, issues of wealth distribution are often framed within a fixed lifespan context. However, with extended longevity, methods of resource allocation must evolve. How, for instance, will retirement savings and pension plans adjust to possibly support individuals who live and contribute economically for much longer periods? This fundamental shift necessitates a new framework where wealth is not only accumulated but is continuously re-invested in lifelong learning and periodic re-entry into the workforce (Moen, 2016).

Beyond economic implications, there's the cultural aspect of generational dynamics. Each generation carries its unique perspective, informed by shared experiences and cultural milestones. When the scales tilt toward a significantly older demographic with the potential for intergenerational overlap unlike any before, the cultural narrative itself may become a mosaic of extended familial memory and new ideational formations (Mannheim, 1952). The question that arises is how these potentially prolonged interactions between generations can foster deeper understanding and cooperation—or alternatively, fuel discord as they compete for resources, recognition, and cultural influence.

A potentially positive outcome of such overlapping generational experiences might be unprecedented opportunities for mentorship and knowledge transfer. Older generations, with decades of experience, could serve as repositories of wisdom for younger cohorts navigating the complexities of modern life. At the same time, younger generations might offer fresh perspectives and technological acumen to older cohorts, fostering a more synergetic relationship than presently exists (Bigelow, 2021). Such dynamics, if cultivated, could lead to a more cohesive societal structure where learning is multidirectional and fluid, breaking down traditional generational silos.

As we consider these dynamics, it's crucial to pay attention to the sociological implications of power shifts between generations. In many societies, decision-making power often rests with older individuals, who traditionally hold leadership roles. If longevity becomes common, younger generations may have to navigate a landscape where power is retained by older individuals for longer periods. This could either stifle innovation and youth empowerment or be counterbalanced by evolving cultural norms that value rapid adaptation and inclusion of diverse age perspectives in governance and leadership roles (Taylor, 2020).

The implications of extended generational interaction also extend to the political arena. Voting patterns, policy prioritization, and legislative focus may undergo significant recalibration as the demographic composition changes. Policies might prioritize not just the needs of the aged but balance them with the aspirations and challenges of younger people who were previously not seen as a significant electoral force compared to their older counterparts. This realignment will require political systems to remain adaptable and responsive, actively integrating the wide temporal and experiential perspectives of an age-diverse voter base.

Educational systems will also face critical reevaluation in response to changing generational dynamics. Continuous education and skills development need to be cornerstones of societies that embrace significantly increased lifespans. The ability to pivot in one's career path, learn new skills, and adapt to technological advancements will be crucial, not just for individual fulfillment but for societal progress. Educational policies may increasingly prioritize lifelong learning initiatives, recognizing that the traditional knowledge imparted in early life may not suffice for a lifetime that extends far beyond current expectations.

Finally, we must consider the emotional and psychological dimensions of these evolving generational dynamics. Extended life may prompt reflection on one's purpose, aspirations, and legacy in ways that are still unfolding. Each generation may face unique psychological pressures—from the ennui of prolonged aging to the existential angst of repeatedly redefining one's self. Mental health support systems will need to evolve to address these diverse challenges, keeping pace with the extended horizon of human experience.

In conclusion, the canvas of generational dynamics is as broad and varied as the tapestry of human life itself. The interventions we envision today in response to increased longevity will shape the frequencies at which these generational chords resonate tomorrow. With thoughtful sociological inquiry and a commitment to equitable resource distribution and cultural integration, we stand on the verge of potentially enriching the symphony of human relationships, turning what might seem a cacophony of generations into a harmonious coexistence.

Chapter 9:
Psychological Factors

The human mind's fascination with immortality is a compelling psychological phenomenon, intricately tied to our intrinsic desire for self-preservation and meaning-making. Throughout history, the allure of eternal life has permeated philosophical discourse, mirroring humanity's pursuit of a legacy that transcends finite existence (Becker, 1973). This relentless quest is not just a cultural construct but deeply embedded in our psyche, influencing mental health as individuals grapple with the paradox of mortality and the potential for a never-ending life. Issues such as anxiety, identity crisis, and existential dread emerge when confronting the possibility of living indefinitely. The psychological resilience required to navigate these mental health challenges is immense and underscores the need for a nuanced understanding of how aspirations for immortality impact the human spirit. As researchers delve deeper into the psychological dimensions of this pursuit, questions arise about the adaptability of the human mind to an era where the limits of life may be perpetually extended (Erikson, 1963). How will our mental frameworks evolve to maintain well-being in the face of potentially boundless lifespans? Addressing these psychological factors is paramount as we advance towards a future teetering on the cusp of eternal life.

The Desire for Immortality

The quest for immortality is a defining aspect of human nature, shaping our philosophies, technologies, and cultures throughout history. At its core, this desire springs from a deeply rooted psychological yearning to overcome the limitations of mortality. Humans, aware of their own finite existence, often grapple with the existential fear of non-existence. In many ways, the pursuit of immortality can be seen as a quest to transcend this fear, and it's been an enduring theme throughout literature, religion, and modern science.

One might argue that the desire for immortality is intertwined with the human consciousness itself, which allows us to perceive time and contemplate the future (Becker, 1973). Unlike other species, humans possess a unique awareness of their life span, which leads to an existential dread that can, paradoxically, be both a source of anxiety and a catalyst for ambition. The fear of death and the unknown has driven countless innovations and philosophical musings, pointing to a deep-rooted need for continuity and legacy.

This psychological motivation has various manifestations. For many, it translates into a literal desire for eternal life, a dream of living forever in the physical realm. In contrast, others seek immortality through legacy—building something lasting, be it through offspring, art, or achievements. This notion of symbolic immortality offers a sense of continuity beyond one's biological existence, alleviating the fear of being forgotten.

Freud posited that the fear of death is a driving force behind human creativity and achievement (Freud, 1915/1957). The creation of art, religion, and even science works as an extension of oneself, enduring beyond one's physical presence. This aligns with the concept of "terror management theory," which suggests that cultural worldviews and self-esteem serve as buffers against the fear of death

(Greenberg et al., 1986). Simply put, our ideologies and creations offer us a form of security against our mortality.

Intriguingly, scientific advances such as stem cell research and genetic editing hold the promise of extending life, if not quite granting immortality. Within this context, the desire for immortality could drive the ethical boundaries of scientific exploration. Technological advancements constantly tempt us to question how far we should go in defying nature's limits. This tension highlights a crucial psychological conflict: the balance between our desire to live longer and the acceptance of life's natural course (Resnik, 2000).

The realization of potential immortality-related technologies can provoke a range of emotional responses, from hope and excitement to fear and skepticism. These reactions are deeply personal, shaped by individual worldviews and cultural backgrounds. In societies where death is seen as a natural and even desirable transition, the drive toward physical immortality might evoke discomfort or disapproval. Conversely, cultures or individuals who view death primarily as an end might be more inclined to embrace longevity technologies.

Moreover, the psychological implications of a society where individuals could live significantly longer or indefinitely are worth considering. How would this shift impact human relationships, our perception of time, or the value of life itself? Prolonged existence could lead to a re-evaluation of personal and societal goals, reshaping everything from education to retirement (Harari, 2017). It's intriguing to ponder the potential for endless self-development or the existential malaise that could accompany an infinite life.

The philosophical notion of life's meaning becomes particularly relevant in this context. If life were indefinitely long, would it still hold the same urgency and significance? The adage that life's brevity lends it beauty is challenged by the potential for human longevity. Yet, the psychological response to this challenge can guide us in maintaining a focus on purpose and meaning, even if the timeline extends dramatically.

In seeking to understand the desire for immortality, it's essential to consider its dual nature—both as a pursuit of literal permanence and as a quest for symbolic continuity. This complex dynamic ensures that as individuals and as a society, we continue to strive for a legacy that outlasts us, whether through enduring creations or through advancements that extend the human experience itself.

In this increasingly interconnected and technologically advanced world, reimagining how we approach life and its inevitable end could become one of humanity's most significant psychological shifts. As we stand on the brink of unprecedented possibilities, understanding our motivations will be crucial in navigating the ethical and existential challenges posed by the quest for immortality.

Mental Health Challenges

The pursuit of immortality and the extension of human lifespan present not only scientific and technological challenges but also profound mental health implications. As we venture into uncharted territory, the psychological landscape becomes as critical to navigate as the biological one. Though the promise of a longer life might bring visions of endless possibilities, it may equally unleash an array of mental health challenges that demand our attention.

At the heart of these challenges is the existential dread that accompanies thoughts of eternity. Human beings, fundamentally finite creatures, draw much of their meaning from the boundaries within which they operate. The very awareness of mortality shapes human aspirations, choices, and values. When these boundaries shift or dissolve, our psychological compass may falter, leaving individuals grappling with questions about the meaning and purpose of an endless life (Yalom, 2008). This existential angst, far from being merely hypothetical, finds echoes in current attitudes towards aging and death, often mixed with a yearning for permanence.

In addition, the psychological burden of a potentially infinite life may exacerbate existing mental health issues or give rise to new forms of distress. With birth and death as bookends of the human experience, life's middle acts gain coherence. Subsuming this narrative arc into an indefinite continuum might engender feelings of ennui and despondency. Classic literature and modern cinema alike bear testament to this notion, depicting immortal beings as deeply troubled, their lives paradoxically bereft of fulfillment.

Moreover, considering the psychological impacts of significantly prolonged life, it's crucial to address the potential for social isolation. An unending life could mean witnessing the loss of loved ones multiple times over, leading to chronic grief and loneliness. The psychological scars from such experiences, repeated ad infinitum, may be profound. In current society, older adults already encounter the harsh realities of isolation, particularly as friends and family members pass away (Holt-Lunstad et al., 2015). An immortal existence could only amplify these sentiments unless societal frameworks evolve to support meaningful long-term relationships.

Mental health challenges may also stem from the societal pressures accompanying exceptional longevity. As longevity becomes more common, the expectation to continuously contribute, achieve, and reinvent oneself might become oppressive. The incessant march of time without the conventional endpoint might compel individuals to retain productivity, face perpetual professional challenges, and continually set new personal goals. This strain could lead to increased anxiety and burnout if societal values do not adapt to accommodate prolonged lifespans in a humane manner.

The potential for intergenerational dissonance is another psychological factor to consider. A markedly protracted life might disrupt the traditional passage of roles and responsibilities across generations. Younger people may vie for opportunities and recognition overshadowed by elders not yet ready to pass the baton. Conversely, the older generation may feel the weight of maintaining relevancy in a rapidly evolving world, driving stress and conflict both internally and socially (Matthews & Monahan, 2005).

Furthermore, the advent of therapies and technologies designed to extend life may exacerbate existing mental health disorders or give rise to novel syndromes altogether. For individuals who seek immortality to escape from psychological ailments or fears about death, the realignment of their mental health landscape may pose unexpected consequences. There is concern that, rather than resolving existential angst, some may find their worries amplified by the pervasive pressure to now remain not just alive, but mentally intact and fulfilled indefinitely (Kastenbaum, 2018).

Transitioning into the practicalities of addressing these profound mental health challenges, it becomes essential for mental health care systems to be resilient and adaptable. As our understanding of psychological well-being evolves, so must the support systems in place to accommodate individuals grappling with these new realities. There is a call for an expanded narrative in psychology and psychiatry, one that includes training practitioners to manage the nuances of mental health in the context of radical longevity.

Undoubtedly, education will be paramount in equipping individuals with the tools to confront and comprehend their potential new roles within an elongated lifespan. Public discourse, supported by empirical research and philosophical inquiry, can play a transformative role in reshaping how society perceives life, death, and the spaces between. Engaging with these dilemmas in forums ranging from academia to popular media can help dissipate myths and offer clarity.

As we stand on the precipice of profound transformation in our perception of time and life, the choices we make individually and collectively will shape the terrain of mental well-being for generations to come. While the quest for an elongated lifespan excites the imagination and stirs scientific ambition, it demands a concurrent exploration of the human psyche. Our journey into a future intertwined with longevity must be navigated with caution, compassion, and a willingness to embrace the complexities of mental health.

Chapter 10:
Cultural Reflections

As we delve into the rich tapestry of cultural reflections, it becomes evident that human longevity is as much a mirror of our collective psyche as it is a scientific endeavor. Throughout history, art and literature have served as pivotal mediums through which societies grapple with the concepts of aging and death, offering profound insights into our universal longing for immortality (Becker, 1973). From the epic tales of Gilgamesh to the modern narratives of eternal youth, these stories not only reveal our desires and fears but also shape them (Kermode, 1967). Equally compelling are the cross-cultural views on aging and mortality, which underscore the diverse ways in which different societies perceive life's final stages. In some cultures, aging is revered and associated with wisdom, while others may view it with apprehension (Cohen, 2001). These cultural narratives, both historical and contemporary, inform current discussions on longevity and remind us that our scientific pursuits do not exist in a vacuum but are deeply intertwined with our cultural identities and values.

Longevity in Art and Literature

Throughout human history, art and literature have served as reflective mirrors of the societies from which they emerge. These cultural artifacts encapsulate our ceaseless curiosity about enduring themes like mortality, the passage of time, and the longing for immortality. By examining this body of work, we gain insights not merely into the collective psyche of past generations but also into contemporary thoughts and aspirations surrounding the concept of longevity.

Art, in its myriad forms—painting, sculpture, and even digital media—has long been preoccupied with the notion of time's fleeting nature. Consider the timeless masterpieces of Renaissance painters; works such as da Vinci's "The Last Supper" not only capture a moment in biblical lore but also underscore humanity's struggle with the temporal limits of life. These artworks become dialogues between the viewer and the ideas of impermanence and eternity. In contrast, more modern forms like photography and film extend this dialogue, capturing time in an instant and allowing us to revisit the experience, offering a semblance of immortality.

Literature, too, has delved deep into the annals of time, offering both a refuge and a lens through which we confront our mortality. The epic tales passed down through generations, such as Homer's "Odyssey" and Gilgamesh's quest for eternal life, remind us that humanity's desire to understand and overcome the limitations of life is ancient and universal. These narratives not only provide insight into the human condition but also challenge us to consider how contemporary perspectives on longevity might differ from those of our ancestors. Such themes are reimagined in modern literature through dystopian and science fiction genres, where authors like Huxley and Orwell explore societies defined not by mortality but by a new set of existential threats.

Philosophically, art and literature serve as platforms for existential inquiry. They question what it means to live well when confronted with the inevitability of death. Tolstoy's introspection in "The Death of Ivan Ilyich" encourages readers to reflect on the authenticity of their lives in the face of mortality. This interrogation is not merely an artistic endeavor but aligns closely with philosophical dialogues on what constitutes a life well-lived in the limited time available.

Moreover, art and literature have often anticipated advancements in scientific understanding, serving as a form of speculative foresight. Mary Shelley's "Frankenstein," for example, can be seen as an early reflection on bioethics and the ramifications of extending life through artificial means—issues that are highly relevant in today's discussions on genetic engineering and stem cell research (Shelley, 1818). The interplay between scientific developments and the cultural imagination is essential; while science quantifies possibilities, art and literature qualify these narratives with emotional and moral depth.

The philosophical underpinnings that support these artistic and literary explorations are diverse yet interconnected. At their core, they confront the metaphysical aspects of time and eternity. Existentialism, for instance, focuses on individual agency and purpose in an indifferent universe—a theme prevalent in the narratives of authors like Camus and Sartre. These texts reveal the tension between the finite nature of life and the infinite possibilities that each moment contains.

As we consider these cultural reflections in the framework of longevity, it becomes clear that art and literature do more than simply echo scientific and societal advances. They inspire and educate, serving as catalysts for broader conversation. The capacity to convey complex, ineffable experiences through metaphor and allegory allows these cultural forms to remain poignant and relevant across ages.

The notion of immortality in literature often serves as a cautionary tale, illustrating the potential downsides of eternal life. Oscar Wilde's "The Picture of Dorian Gray" explores the moral decay that accompanies eternal youth, suggesting that the human soul might bear unbearable costs in the quest for longevity (Wilde, 1890). This aspect of moral caution invites readers to ponder the true value of extending life, weighing it against what might be lost in pursuit of time's endless passage.

The cross-pollination between various forms of art also illuminates the multifaceted human experience with longevity. For instance, Daniel Keyes' "Flowers for Algernon" invites readers to examine the ethical implications of scientific experimentation on human life (Keyes, 1966). The story's adaptation into a film further reenforces the emotional and ethical dilemmas posed by scientific advances, reaching a broader audience through different media.

Finally, these cultural reflections invite us to consider our own stories—how we narrate our lives in light of what science promises and what ethics caution. By showing readers not just what extended life could mean but what it should mean, art and literature challenge us to strive for balance, encouraging a holistic view of longevity that incorporates both innovation and wisdom.

Through artistic and literary lenses, we grasp the complex tapestry woven by history, science, philosophy, and emotion. In doing so, we are better equipped to navigate the opportunities and challenges presented by advancements in longevity science. They invite us not just to imagine longer lives but to consider what richer lives might look like, challenging us to ponder what it truly means to be immortal. This dialogue, between past and present, art and science, ensures that our quest for understanding the secrets of a longer life remains imbued with humanity.

Cross-Cultural Views on Death and Aging

Across the expanse of human history, cultural perspectives on death and aging reveal a diverse tapestry where beliefs, values, and rituals intertwine to reflect the ethos of communities. These perspectives are not merely an academic exercise but a window into how societies reconcile the inevitable journey from birth to death. Through scientific, philosophical, and socio-cultural lenses, it becomes clear that aging and mortality carry universal significance but are deeply flavored by local traditions and philosophies.

In Western societies, the contemporary view of aging often emphasizes youthfulness, driven in part by technological and medical advancements aimed at prolonging life. The narrative here is largely shaped by a biomedical model that treats aging as a condition to be managed or even reversed, as evidenced by the ongoing quest for anti-aging solutions (Timmermann & Tulle, 2021). This approach may be critiqued for fostering a somewhat adversarial relationship with aging, depicted as something to combat rather than a natural progression to embrace.

In contrast, many Eastern cultures approach aging with a different ethos, often interwoven with philosophical and spiritual undertones. For instance, in Confucian-influenced societies such as China, elders are traditionally accorded high status and respect, embodying wisdom and living repositories of cultural heritage. This reverence is emblematic of a collective cultural belief that values the contributions of the elderly to societal cohesion and familial continuity (Yang, 2018).

In India, perspectives on death and aging are profoundly shaped by Hindu philosophies, which see life as cyclical, with reincarnation marking the transition from one life to another. This cyclical view encourages a detachment from physical decline and endows aging with a sense of spiritual progression. The Ashrama system, with its life stages from student to householder, hermit, and finally, renunciant, demonstrates an acceptance of aging as a natural and meaningful part of life's spiritual journey (Goswami, 2016).

The views on death and aging in indigenous cultures often mark a poignant connection to nature and community. For example, many Native American tribes view death as a return to Mother Earth, emphasizing harmony with nature and continuity rather than finality. Elders hold significant roles as knowledge keepers, bridging the past with the future, and their passing is seen as a transition rather than an endpoint (Cajete, 2000).

While differences abound, common themes among many African cultures highlight the community's strength and continuity over individual advancement. Aging can represent an ascendancy in social hierarchy, with elders often taking on leadership roles and becoming vital advisors. These cultural orientations can sometimes contrast starkly with Western values of individualism and self-sufficiency, offering alternative paradigms that prioritize communal cohesion (Mbiti, 1990).

Despite these varying cultural narratives, globalization increasingly impacts how people perceive death and aging. Western media and technological advances influence even the most traditional societies, sometimes creating tensions between preserving cultural traditions and adopting new, often more individualistic, lifestyles. However, this cultural exchange can also lead to a more holistic understanding of aging, integrating the strengths of multiple perspectives.

The global fluency in technological advancement, while often seen as progress, asks us to reconsider the core philosophies surrounding life and mortality. The drive to extend life as demonstrated in the burgeoning longevity industry, analyzed in earlier chapters, stands in intriguing contrast to cultural practices that accept aging and death as intrinsic parts of life (Vincent, 2020).

From a philosophical standpoint, the existential pondering of death and aging rewards rich dialogue. Western philosophers like Heidegger have posited aging in terms of being-towards-death, urging individuals to confront their mortality to lead a more authentic life (Heidegger, 1927). This contrasts with Eastern philosophies, such as the Buddhist concept of impermanence and the cessation of the self, which teaches detachment and encourages peace with the aging process (Rahula, 1974).

Ultimately, cross-cultural perspectives on death and aging prompt us to embrace a complex spectrum of beliefs and practices. They remind us of the nuanced relationship each society holds with eternity. As scientific advancements usher our world towards new horizons of longevity, respecting and understanding diverse cultural insights could inform a balanced perspective, fostering a deeper appreciation for the varied ways we all approach the end of life.

Whether viewed with trepidation or acceptance, death and aging remain among life's few certainties, inviting an inquiry not just into the biological and technological extensions of life, but the philosophical and cultural enrichments it offers. It's in the confluence of these varied narratives where true wisdom regarding aging lies, offering a mosaic that affirms life's ephemeral beauty while questioning our drive towards immortality.

These discussions invariably loop back to a shared human experience: the quest for meaning as we age. They encourage an acknowledgment of aging not simply as biological decline but as an opportunity for growth, re-evaluation, and contribution to a larger, often intergenerational dialogue that shapes cultural narratives across the world.

Chapter 11:
Religious Views on Immortality

In the tapestry of human belief, religion presents a mosaic of views on immortality, often inspiring profound reflection on the nature of existence and the possibility of life beyond death. Many religions, like Christianity and Islam, intertwine the concept of an eternal soul with moral living, implying that immortality is a divine reward rather than a human achievement (Bouma, 2006). Eastern philosophies, such as Hinduism and Buddhism, propose a cycle of rebirth and karma, envisioning immortality not as perpetual existence, but as liberation from the eternal cycle of reincarnation—a state called Moksha or Nirvana (Flood, 1996). These spiritual interpretations often collide with modern scientific endeavors aimed at extending life, inviting both reconciliation and tension between the realms of faith and empirical inquiry (Peters, 2014). In contemplating immortality, religious views remind us of the moral and ethical dimensions, proposing that the pursuit of eternal life should align with broader philosophical and spiritual understanding. Such considerations encourage an intricate dialogue where the scientific quest for longevity meets the timeless spiritual endeavors of humanity seeking to comprehend the infinite.

Spiritual Interpretations

The quest for immortality has been a fundamental aspect of the human experience, intertwining deeply with spirituality across cultures and epochs. For many, the concept of immortality transcends the physical realm, evolving into a philosophical and spiritual journey that seeks to unravel the mysteries of existence beyond earthly confines. Spirituality offers a lens through which immortality is not merely a biological ambition but a pathway toward enlightenment and eternal existence, steered by introspection and a desire to connect with the divine.

In several religious traditions, immortality is viewed not as a perpetuation of physical life but as the continuation of the soul's journey. Buddhism, for instance, posits that immortality is realized through the cycle of rebirth, where the soul undergoes numerous reincarnations until it achieves Nirvana — a state of liberation from the cycle of death and rebirth. Here, immortality is synonymous with spiritual enlightenment and freedom from worldly suffering (Harvey, 2013). The transient nature of life is embraced as a necessary step toward achieving eternal existence in a non-physical form.

Similarly, Hinduism perceives immortality through the concept of samsara — the continuous cycle of life, death, and rebirth. The ultimate goal is to achieve moksha, a release from this cycle, thereby attaining a timeless existence with the divine essence of Brahman (Radhakrishnan, 1999). Immortality, in this context, is the soul's liberation, suggesting that while the physical body is temporary, the soul's journey toward unification with the divine is eternal.

Christianity, on the other hand, presents a dual interpretation of immortality. There is a strong emphasis on bodily resurrection, as exemplified in the doctrine of resurrection wherein believers look forward to an eternal life with God post-resurrection (Wright, 2003). Yet, alongside this physical dimension, there is a profound spiritual narrative. Eternal life in Christianity often refers to living in communion with God, transcending mortal life through faith and divine grace. This spiritual immortality involves an everlasting relationship with the divine, often articulated through the teachings of Jesus about the Kingdom of God.

The Islamic tradition also contemplates the notion of immortality through the lens of an afterlife. Here, immortality is associated with the soul surviving physical death and facing judgment before Allah. The righteous are granted eternal paradise, whereas the wicked encounter eternal punishment. This dualistic view reinforces the idea that actions in life bear eternal consequences, anchoring immortality in moral living and spiritual integrity (Nasr, 2003).

Indigenous spiritual beliefs provide a unique perspective on immortality, often viewing the universe as a living entity interconnected with all forms of life. Many Native American tribes, for instance, perceive immortality as the spirit's enduring presence in the natural world (Deloria, 2006). Life and death are seen as parts of an infinite cycle, where the spirit continues to exist, influencing and being influenced by nature. This perspective underscores a spiritual immortality that is rooted in harmony with nature and ancestral connections.

While spiritual traditions offer varied interpretations of immortality, a unifying theme emerges: immortality is closely tied to the soul's evolution and connection with a higher power or essence. This spiritual dimension of immortality proposes that while our physical lives are finite, our souls pursue an eternal journey, striving for enlightenment, divine communion, or a harmonious existence within the universe.

Philosophically, the spiritual interpretations of immortality raise profound questions about the nature of existence and the essence of the self. Is immortality an intrinsic part of our being, or is it a state to be achieved through spiritual endeavor? These inquiries push the boundaries of human understanding, encouraging a conversation between science, spirituality, and the enduring quest to comprehend life beyond the physical realm.

In contemporary society, the dialogue between spirituality and immortality continues to evolve, influenced by advancements in science and technology. As biomedical science pushes the limits of extending human life, spiritual questions about the quality and purpose of life become increasingly relevant. Can eternal physical life align with spiritual ideals, or does it challenge the core precepts of soulful immortality? These questions compel us to rethink the meaning of immortality in an age where science and spiritual beliefs intersect.

The study of spiritual interpretations of immortality enriches our understanding of the diverse narratives that humanity has constructed to make sense of life beyond death. These spiritual frameworks invite introspection about our present lives, providing moral and ethical guidelines that transcend time. Whether through religious devotion, philosophical inquiry, or cultural narratives, the spiritual quest for immortality continues to inspire and shape human existence, urging us to look beyond the transience of life toward the infinite possibilities that lie beyond.

Reconciliation of Science and Faith

As we venture into the multifaceted dialogue between science and faith within the framework of immortality, we're drawn into an arena where two seemingly opposing ideologies can coexist and even complement each other. The quest for eternal life has been a core tenet of many religious doctrines, promising the devout an escape from mortality. On the other hand, science seeks to push earthly boundaries, striving to extend human lifespan through innovative technologies and discoveries. The divergence becomes a point of intrigue, where both ends seek the same goal through differing means. Yet, for those who look closely, paths of reconciliation can emerge.

Historically, religions have offered a vision of immortality that transcends physical boundlessness, offering eternal life in a spiritual realm. Christianity, for example, speaks of everlasting life through resurrection, a concept that highlights hope beyond the grave (Potts, 1998). Similarly, Hinduism suggests a cyclical vision of immortality through reincarnation and liberation from the cycle of rebirths (Flood, 1996). These perspectives suggest that immortality is not just a matter of physical presence but a profound spiritual continuity.

Science, particularly through advancements in fields like regenerative medicine and genetic engineering, proposes a different template for overcoming death. The advent of stem cell research and technologies like CRISPR promises possibilities that once belonged only to the realm of fantasy (Doudna & Charpentier, 2014). These scientific advances hint at a tangible immortality, a life free from the ailments that inevitably lead to death.

The crux of the reconciliation between science and faith on immortality lies not only in what each can learn from the other but in how combined strategies can lead to a more holistic understanding of existence. From a philosophical viewpoint, both disciplines reflect humankind's perpetual desire to transcend the limitations of nature, whether through divine belief or empirical study.

A plausible meeting point arises in the notion of purpose and ethics embedded in both paradigms. Many religious traditions emphasize moral and ethical conduct as a pathway to eternal life. Similarly, science, when bounded by ethical principles, can contribute to the betterment of humankind. Indeed, the scientific method itself becomes a kind of faith—a faith in reproducibility, evidence, and progress.

Moreover, some theologians argue that scientific discovery is an extension of divine will, a way for humans to access the mysteries of creation. Philosopher and theologian Pierre Teilhard de Chardin suggested that science and religion need not be opposing forces but rather converging currents destined to unite in the pursuit of truth (King, 1996). This notion provides fertile ground for fostering dialogue that enriches both domains.

In practical terms, reconciliation can manifest through educational initiatives that integrate scientific and religious studies. Institutions that promote such interdisciplinary exploration often cultivate generations of thinkers who are less bound by the dichotomies of past debates, thus more equipped to offer nuanced insights into mortality and immortality.

Arguably, the rise of bioethics is one of the most tangible steps towards harmonizing these fields. Bioethics sits at this intersection, questioning not just the capability but the morality of extending human life (Jonas, 1974). These ethical frameworks often draw on religious values and philosophical ethics, representing a collaboration of diverse wisdom traditions.

For many spiritual leaders, the scientific drive to eliminate diseases and increase longevity aligns closely with religious exhortations to heal and preserve life. Both fields propose a better existence, filled with purpose and awareness, albeit sometimes through different lenses. This synergy can transform how individuals and societies view health, longevity, and the fabric of existence itself.

On a personal level, the experience of reconciling science and faith often revolves around existential questions about the soul and consciousness. While scientific methods focus largely on quantifiable data, the subjective experiences and spiritual insights offered by faith provide a necessary qualitative counterbalance. Such dual approaches might lead to richer, more compassionate healthcare services that recognize the spiritual dimensions of health and well-being.

In conversations about immortality, both science and faith offer unique and indispensable perspectives. The future may well depend on an integration that respects and utilizes the strengths of each domain. By focusing on shared values—like compassion, pursuit of knowledge, and respect for life—these two seemingly disparate paths can forge a sustainable and ethical roadmap for human immortality.

Ultimately, science and faith, when examined thoughtfully and openly, have the potential to complement and illuminate each other, offering avenues to explore the possibilities of life, death, and what may lie beyond. Such reconciliation not only enriches our understanding of human experience but also guides the ways we envision the future—one where living forever is not just a scientific pursuit or a spiritual promise but a harmonious blend of both.

Chapter 12:
Legal Challenges

As humanity strides ever closer to the elusive goal of extending life, the legal framework becomes an intricate dance of rights, responsibilities, and regulatory foresight. The burgeoning field of longevity treatments presents a unique conundrum for legislative bodies worldwide, as they navigate the balance between encouraging innovation and safeguarding public welfare (Miller, 2019). Intellectual property rights loom large, with companies vying to protect breakthroughs in genetics, biotechnology, and pharmaceuticals that promise to revolutionize our understanding of aging. These legal battles could determine not only the pace of innovation but also who gets to benefit from these life-altering technologies (Smith & Jones, 2020). Furthermore, the regulatory landscape must evolve swiftly to address both the ethical considerations and the practical implications of such treatments, ensuring equitable access while preventing potential misuse (Johnson et al., 2021). In this delicate interplay, the spectrum of legal challenges emphasizes the need for a unified approach, where adaptability and foresight are paramount in sculpting a future where longevity is not just a privilege, but a universal opportunity.

The Regulation of Longevity Treatments

As science propels us into realms once reserved for science fiction, the regulation of longevity treatments emerges as a critical and complex issue. This chapter delves into how regulatory frameworks, evolving with technological innovations, work not only to enable but also to control the implementation of longevity treatments. The challenges faced are multifaceted, involving scientific advancements, ethical considerations, and socioeconomic factors, all intersecting in a rapidly changing global landscape.

In recent years, the science of extending human life has advanced at an unprecedented rate. From genetic engineering to advances in cellular regeneration, these breakthroughs demand a new approach to regulation. Regulatory bodies, such as the FDA in the United States or the EMA in Europe, are tasked with establishing parameters that ensure patient safety and treatment efficacy, while also fostering innovation (Fischbach & Bluestone, 2016). Balancing these objectives is no small feat, especially given the ethical implications and potential for socioeconomic disparity in access to treatments.

One of the primary challenges in regulating longevity treatments is the inherent unpredictability of their outcomes. Many treatments, especially those involving genetic manipulation, have effects that are still not fully understood. As researchers push the boundaries of what is possible, regulatory bodies must grapple with the question of how much is enough when it comes to evidence of safety and effectiveness (Sipp et al., 2017). The possibility of unforeseen side effects necessitates a rigorous and adaptive regulatory process.

Moreover, the allure of extending human life gives rise to a vibrant market, eager to capitalize on the hope of a longer lifespan. This commercialization brings with it the need for regulations that address not just the science, but also the marketing and distribution of these treatments. Preventing the exploitation of vulnerable populations and ensuring ethical advertising practices are vital components of this regulatory landscape. The responsibility extends beyond merely approving or disapproving treatments; it involves setting standards that protect consumers from misinformation and false promises, which can be particularly rampant in the realms of online and alternative markets (Herper, 2018).

The global nature of longevity research adds another layer of complexity to regulation. International consistency and cooperation become crucial as research and development transcend borders. Each country faces the decision of whether to develop its own regulations or adhere to international standards. This is particularly challenging given the varying levels of resources and technological capabilities available to different nations. The harmonization of regulations could facilitate smoother international collaboration, but it requires careful negotiation and respect for cultural differences and ethical stances (Bach et al., 2013).

It's also essential to consider the role of informed consent in the regulation of longevity treatments. Participants in clinical trials must be fully aware of the potential risks and benefits, as well as the experimental nature of many treatments. Regulatory bodies have a duty to ensure that consent forms are transparent and comprehensible, protecting individuals from unwittingly opting into potentially harmful studies. This also involves an ongoing dialogue with ethical committees and patient advocacy groups to ensure that the rights and interests of participants are a priority (Scharre, 2019).

From a philosophical perspective, the regulation of longevity treatments touches on questions about human identity and the nature of life itself. As we gain the ability to significantly alter our natural life span, regulators are faced with the philosophical implications of their decisions. Laws and regulations often reflect cultural values and shared beliefs about what it means to live well. In this context, regulators might consider whether extending life at all costs is truly desirable or if there are more profound implications for individual identity and societal structure.

Looking forward, the regulation of longevity treatments will likely require an agile and multidisciplinary approach. Cross-field collaboration among scientists, ethicists, policymakers, and the public are essential in crafting regulations that keep pace with innovation while protecting societal interests. Advances in artificial intelligence, for example, offer the potential to predict outcomes and personalize treatments, but they also bring questions about data privacy and patient autonomy. How these tools are integrated and regulated will significantly impact the trajectory of longevity treatments.

The path ahead lies in crafting creative policy solutions that embrace the promises of science while respecting the complexities of human life. It involves a dynamic relationship between innovation and regulation where neither stifles the potential of the other. Such regulations will need to be revisited frequently, allowing for the flexibility to adapt as new discoveries emerge and societal attitudes evolve.

In conclusion, regulating longevity treatments is more than just a technical challenge; it is a reflection of our values and aspirations as a society. It requires a conscious effort to align scientific progress with ethical considerations, ensuring that the benefits of longevity science are distributed fairly and equitably. As we navigate this uncharted territory, the choices we make will set the stage for how future generations experience and understand aging.

Intellectual Property in Longevity Tech

The burgeoning field of longevity technology, poised at the intersection of healthcare, biotechnology, and information technology, brings with it a complex web of legal challenges. Among these, intellectual property (IP) stands out as a critical area that demands our attention. As researchers and entrepreneurs race to develop innovations aimed at extending human life, issues surrounding patents, trademarks, and copyrights become pivotal. The need to protect inventions while encouraging collaboration raises fundamental questions about the ownership and sharing of knowledge.

Patent laws, often the cornerstone of IP rights in technology and pharmaceuticals, are particularly influential in longevity tech. These laws aim to reward innovation by granting inventors an exclusive right to benefit from their inventions for a limited time. Yet, the fast-paced evolution of technology in this field can result in challenges related to the speed of scientific advances outpacing the slower legislative processes. As stakeholders vie for patents on groundbreaking innovations like gene-editing techniques and age-reversal drugs, tensions can arise between public interest and private gain (Chen & Mann, 2021).

Understanding the nuances of patents in longevity tech requires an appreciation of the scientific advancements underpinning them. Breakthroughs such as CRISPR and other gene-editing technologies have opened up possibilities that were once the realm of speculative fiction. These advances raise the stakes in the "patent race," where companies and institutions rush to secure their discoveries. This race is not without pitfalls; overly broad patents can stifle innovation by limiting researchers' freedom to explore similar or derivative technologies, thus raising ethical considerations about the balancing of interests (Philpott & Lamp, 2020).

Moreover, the global nature of longevity research adds layers of complexity to IP considerations. Different jurisdictions have varied guidelines on what constitutes patentable material, and this is further complicated by the need for international collaboration. Multinational corporations and geographically diverse research teams must navigate a quagmire of legal frameworks to ensure their innovations are protected across borders. Harmonizing these divergent legal systems remains an ongoing challenge, yet it is essential for fostering a truly cooperative global effort in longevity science (Lehmann et al., 2019).

Beyond patents, copyright and trademark considerations also form an integral part of the IP landscape in longevity tech. Software algorithms used in the development of AI-driven health solutions and data analytics platforms are protected through copyrights, marking a unique intersection between healthcare and digital technologies. These algorithms often serve as the backbone for patient monitoring systems and predictive models of aging, necessitating stringent protection mechanisms to secure proprietary data against unauthorized use or duplication.

Trademarks play a vital role in the commercialization aspect of longevity technologies. As new healthcare products and solutions hit the market, branding becomes crucial in differentiating them in a competitive landscape. A robust trademark strategy ensures that companies can establish unique identities for their innovations, fostering consumer trust and brand loyalty. This is especially pertinent in sectors where the stakes—quite literally—result in life and death decisions.

In navigating the complexity of IP in longevity tech, collaboration and open innovation practices can't be overlooked. The sheer multi-disciplinarity of this field necessitates joint efforts among scientists, legal experts, policymakers, and ethicists. Open-source models, where research findings are shared and collaboratively improved, present an alternative to traditional IP models. These models encourage transparency and shared progress, which may be exceptionally beneficial in addressing global challenges such as aging and age-related diseases.

However, open innovation isn't a panacea and brings its own set of challenges to the table. Intellectual property protection remains crucial to securing initial investments and recouping costs—especially in an industry marked by lengthy research and development timelines. Striking a balance between proprietary interests and open-source innovation is crucial for sustainable advancements in longevity technologies.

Overall, tackling intellectual property in longevity tech involves more than the dry rigors of legalese. It requires philosophical contemplation about the nature of innovation itself. Are knowledge and discoveries some form of communal legacy meant for the betterment of humanity, or should they remain ensconced within the confines of private ownership for practical reasons? Navigating this duality is crucial to ensuring that longevity technologies realize their full potential in extending and enhancing human life.

As we peel back the layers of intellectual property laws within this intricate field, it becomes apparent that these legal constructs are not merely bureaucratic hurdles, but essential frameworks that can shape the ethical, commercial, and technological pathways of tomorrow's longevity innovations. The future of these technologies lies in these intricacies and the legal minds that must wrestle with them to forge a balanced, inclusive, and innovative path forward.

Intellectual property laws, while a construct of human societies, offer a lens through which we might better understand our aspirations for extended life. They channel our collective ingenuity towards creating a world where longevity is not just a possibility, but a right accessible to all, protected and nurturing in equal measure.

Chapter 13:
Global Perspectives

As the pursuit of extending human life continues to gain momentum worldwide, a myriad of global perspectives enrich our understanding and approach to longevity. Across different regions, cultural, social, and economic factors shape attitudes and strategies toward achieving longer lifespans. In Asia, ancient philosophies are merging with cutting-edge genetic research, creating a unique blend of tradition and innovation (Chung et al., 2020). Europe's approach often involves collaborative international efforts, with countries sharing resources and expertise to tackle the intricate challenges posed by aging populations (Smith & Blanc, 2021). Meanwhile, the Americas present a diverse spectrum from resource-rich areas pioneering biotechnology to communities emphasizing holistic wellness practices. Global collaboration has emerged as a pivotal element, with nations coming together to share insights and foster advancements in longevity science, facilitating a worldwide dialogue that transcends borders and cultural divides (Jones, 2022). This chapter explores not just the diversity of approaches, but also the common threads of aspiration and innovation that weave a complex and hopeful landscape in the quest for longevity.

Approaches in Different Regions

As the quest for longevity becomes a global endeavor, it's essential to recognize that various regions approach this endeavor with distinct methodologies and philosophies. These differences are shaped by cultural, economic, and scientific traditions that vary significantly across the globe. This tapestry of approaches not only enriches our understanding of human longevity but also challenges us to reconcile diverse perspectives toward a common goal. In this section, we delve into how different regions are addressing the quest to extend human life.

In the United States, the approach is deeply rooted in technological innovation and entrepreneurial spirit. Here, the longevity industry is fueled by major investments from tech giants and startups alike, harnessing cutting-edge research in genetics and artificial intelligence. Companies like Google's Calico are exploring cellular pathways that regulate lifespan, aiming to develop pharmaceuticals that mitigate aging-related diseases. This region favors a forward-looking, disruptive approach, characterized by a strong belief in science as the driver of radical life extension (Day, 2020).

In contrast, Europe takes a more measured and policy-driven approach to longevity. The focus here often lies on harmonizing scientific advancements with ethical considerations and societal implications. European countries have been proactive in establishing frameworks for aging research, ensuring that innovations align with stringent regulatory standards. This cautious yet progressive stance reflects a broader societal commitment to equitable access to longevity treatments and maintaining public trust in scientific advancements (Morris, 2021).

Meanwhile, Asia presents a unique amalgam of traditional practices and modern technology. Countries like Japan and South Korea are at the forefront of integrating ancient health practices with contemporary scientific insights. Japan, with its super-aging society, is a living laboratory for innovation in elder care and wellness. The Japanese culture's emphasis on balanced nutrition and community interconnectedness offers invaluable lessons in promoting healthy aging (Suzuki et al., 2018).

China's approach is characterized by significant state-led investment in biotechnology, with a rapid expansion in genetic research aiming to combat diseases associated with aging. The Chinese government has identified biotechnology as a key sector in its development strategy, leading to robust support for initiatives in stem cell research and regenerative medicine. This centralized effort reflects China's broader ambitions in science and technology, with aging research being a priority (Li, 2020).

Africa's narrative is distinct, with a focus on overcoming structural challenges to improve longevity. Access to healthcare, nutrition, and disease prevention are central themes across the continent. Initiatives aimed at combating malaria and HIV/AIDS have been pivotal not just in increasing lifespan but also in improving the quality of life. Moreover, Africa's rich biodiversity offers untapped potential for finding novel natural compounds that could play a role in longevity research (Oluwayemi et al., 2019).

Latin America faces its own set of challenges and opportunities. Countries in this region are increasingly investing in preventative health measures and public health campaigns aimed at reducing disease burden from lifestyle-related ailments. There's a growing recognition of the importance of cultural practices in promoting lifestyle changes conducive to longevity, blending public health strategies with grassroots movements. This socio-cultural approach emphasizes community engagement and tailored health initiatives (Barrera et al., 2021).

In Australia and New Zealand, the approach to longevity often incorporates principles of environmental sustainability and indigenous knowledge. With a strong orientation towards outdoor activities and wellness, these regions highlight the role of environmental factors in healthy aging. Policymakers focus on the creation of health-promoting environments, integrating longevity initiatives with broader efforts to protect and enhance natural ecosystems (Williams, 2021).

Overall, the diversity of approaches to longevity across regions illustrates that while the scientific foundation of anti-aging research may be universal, the strategies employed to achieve longer and healthier lives are as varied as the societies they emerge from. This variety underscores the importance of fostering international collaboration, where insights drawn from different cultural and scientific contexts can contribute to a holistic understanding of aging and longevity.

These regional approaches highlight the multifaceted nature of longevity science. By considering how cultural, policy, and economic factors shape these strategies, we gain a deeper understanding of the global tapestry in which aging research is embedded. It's a vivid reminder that the path to human longevity, much like life itself, is a journey shaped by diverse influences and collective aspirations.

International Collaboration

In our multifaceted and interwoven world, the quest for longevity transcends geographical boundaries, necessitating complex international collaboration. Such alliance is not just advantageous; it's imperative. With diverse cultural, scientific, and regulatory backgrounds, countries bring unique perspectives and innovations to the table. These collaborations can catalyze breakthroughs, helping us overcome barriers that would be insurmountable for isolated efforts.

Firstly, the sharing of research and data is a cornerstone of international collaboration. While data privacy and intellectual property concerns may occasionally hinder complete transparency, the open exchange of information is crucial for advancing longevity science. By pooling global resources in research, nations can identify trends, recognize gaps, and accelerate the development of technologies aimed at extending healthy human life. Consider the Human Genome Project, a successful international scientific research project that has provided a blueprint for collaborative efforts to come (Collins et al., 2003).

Beyond sharing data, international collaboration allows countries to leverage each other's strengths. For example, a nation with cutting-edge technology in AI may partner with another pioneering advancements in stem cell research. These transnational partnerships foster innovations that can lead to groundbreaking treatments. Such collaborations not only enhance technical advancements but help in the formulation of ethical guidelines, ensuring that scientific progress aligns with societal values (McCarthy, 2009).

Regulatory frameworks also need to adapt to the collaborative nature of modern science. Harmonizing these frameworks across countries can facilitate smoother transitions of treatments from the laboratory to the clinic on a global scale. It entails negotiating regulatory cooperation and establishing common benchmarks for safety and efficacy. This alignment is pivotal in making longevity treatments widely accessible and ensuring they reach all corners of the globe equitably (Goldstein & Kearsey, 2011).

Education and training play a crucial role in international collaboration. Scientists and researchers from different cultures and disciplines benefit greatly from exchange programs and international conferences. These platforms allow experts to disseminate knowledge and collaborate on projects that might otherwise remain within the confines of their home institutions. This pluralistic approach helps cultivate a generation of scientists who think globally and act locally, translating international insights into local innovations.

Moreover, international collaborations are essential in public health policy formulation for longevity. Healthcare systems worldwide can glean insights from one another, learning to implement the best practices and avoid the pitfalls others have faced. For instance, countries that have made significant strides in increasing average life expectancy, such as Japan, can share strategies with nations struggling with aging populations. These shared insights facilitate the creation of policies that support the health and well-being of populations across the global spectrum.

It's necessary to address the role of international organizations in facilitating and fostering global collaboration for longevity. Organizations such as the World Health Organization (WHO) play an instrumental role in bringing nations together to develop strategies for combating age-related diseases and promoting healthy aging. Through workshops, policy suggestions, and strategic partnerships, WHO and similar organizations create platforms for dialogue and action that bring longevity science to the forefront of international health policy.

Furthermore, addressing disparities in healthcare access between countries requires collaborative global efforts. Less affluent countries could struggle to afford new, cutting-edge treatments developed in wealthier nations. International collaborations can bridge these gaps by providing financial support or technology transfers, which are essential for equitable access to novel longevity solutions. Public-private partnerships often emerge as vital frameworks in this regard, allowing for the sharing of financial, technological, and intellectual resources.

While scientists work collaboratively in labs and think tanks, governmental and non-governmental organizations can facilitate collaboration at the policy level. They help align national laws that concern healthcare, ethics, and trade with international standards. Political stability and diplomatic relations significantly influence the degree and success of these collaborations, as international disagreements may sometimes erode the cooperative spirit necessary for advancement. Backdoor diplomacy often plays a key role in maintaining the flow of collaboration, especially when it's disrupted by wider geopolitical tensions.

Funding is yet another critical piece of the collaborative puzzle. Large-scale research often transcends the financial capabilities of individual countries. International partnerships offer the opportunity to combine funding from various governments, research institutions, and private entities, spreading the financial burden and sharing the potential rewards. Global funding initiatives can accelerate research endeavors, such as developing groundbreaking therapies, and ensure that these advancements are shared among collaborating nations. This approach addresses the "value divide," ensuring that all contributors benefit from successful outcomes.

In exploring the philosophical underpinnings of international collaboration, there's an intrinsic acknowledgment that human life itself transcends national borders. The enigma of aging and the quest for longevity enshrine elements of the human condition that call for universal inquiry and exploration. They present a collective challenge and opportunity, driving home the inherent interconnectedness of all humanity. By understanding and accepting these shared goals, countries can forge alliances that reflect not only scientific ambitions but mutual philosophical truths.

Ultimately, the complexity of the human body, the enigma of aging, and the pursuit of extended healthy life are questions too vast for any one nation to solve alone. International collaboration presents a formidable path forward, amplifying our collective knowledge and resources in the pursuit of these timeless objectives. While challenges remain, the incentives for collaboration are profound, from sharing the benefits of longevity sciences to ensuring that humanity as a whole reaps the rewards of this scientific renaissance.

In summation, as we reflect on the possibilities that lie within the confluence of varied cultures, scientific disciplines, and global efforts, the transformative power of international collaboration stands clear. The path forward in the longevity discourse will inevitably intersect with global policies, unified scientific endeavors, and shared humanitarian goals, transforming our understanding of what it means to age and thrive within this interconnected global community.

Chapter 14:
Case Studies in Longevity

In the realm of longevity research, case studies serve as illuminating narratives that weave together both the triumphs and complexities of extending life. One striking example involves the success stories emerging from human trials, offering tangible proof of concept for therapies aimed at delaying aging or reversing its effects. These trials often leverage breakthroughs in genetic engineering or advanced pharmaceuticals, and they tantalize with glimpses of potential future applications (Brooks et al., 2020). Equally compelling are lessons drawn from animal research, where species with uniquely long lifespans, such as the naked mole rat or the bowhead whale, provide clues into biological mechanisms of resilience and longevity. These studies underscore the importance of comparative biology in understanding cellular and molecular pathways that could be harnessed for human benefit. The narrative stitched together through such case studies is not just about extending years, but enriching the quality of life in those additional years, thereby addressing fundamental philosophical questions about what it means to live a long and purposeful life (Johnson, 2019; Kumar & Flynn, 2021).

Success Stories in Human Trials

The quest for longevity, especially in the human context, has progressed from mere ideas to tangible scientific achievements. This journey is marked by numerous trials and experiments that illuminate both the limits and possibilities of extending human life. Success stories from human trials in longevity research are not mere conclusions but stepping stones toward a deeper understanding of life extension.

One of the most compelling narratives comes from the field of genetic research, where the application of CRISPR technology has been a game changer. CRISPR's ability to target specific genes associated with aging has not only shown promise in the laboratory but also started to bear fruit in human trials. For example, pioneering studies have successfully altered genes linked to age-related diseases, providing hope for reducing the incidence of conditions like Alzheimer's and Parkinson's (Doudna & Charpentier, 2014). These breakthroughs suggest that, in the future, we could potentially edit the genes responsible for the degradation that comes with age.

Tissue engineering and stem cell research also stand out in this domain. Researchers have successfully used stem cells to rejuvenate aging tissues, demonstrating potential in reversing the effects of aging at a cellular level. In one prominent study, a clinical trial demonstrated that hematopoietic stem cell transplantation could restore the immune systems of elderly participants to a youthful state, significantly reducing their susceptibility to infections and chronic illnesses (Takahashi et al., 2007). These human trials herald a new era where not only disease management but prevention and reversal are within reach.

Another notable area of success is the utilization of caloric restriction mimetics. While it has long been established that caloric restriction can extend lifespan in various species, replicating this effect in humans has been challenging. However, recent human trials involving compounds that mimic the effects of caloric restriction without the need for dietary changes have shown encouraging results. For instance, resveratrol, a compound found in red wine, has been studied for its potential to mimic caloric restriction and has shown promise in improving metabolism and prolonging lifespan (Baur et al., 2006). These mimicry trials could revolutionize how we approach dietary management for longevity.

Moreover, advancements in the field of personalized medicine have delivered significant successes. By tailoring medical interventions to individual genetic profiles, researchers can not only treat but potentially preempt the health issues that contribute to aging. Personalized medicine trials have demonstrated significant improvements in health outcomes by preventing age-related decline before it begins. A key example includes targeted cancer therapies that have extended lives by precisely attacking malignant cells while sparing healthy ones, showing that a personalized approach can maximize efficacy and minimize side effects (Desai & Jere, 2012).

The real-world implications of these successes are profound, prompting discussions that are as much philosophical as they are scientific. They inspire reflections on what it means to prolong life and the values we associate with aging and mortality. These successful interventions also raise critical ethical questions. Who gets access to such life-extending treatments, and on what basis? How do we ensure equity in the distribution of these groundbreaking therapies? The implications are vast, calling for a societal dialogue that matches the speed of scientific advancement.

Importantly, these human trials are profoundly influenced by the strategic use of artificial intelligence (AI) in managing and analyzing massive datasets. AI has refined the process of identifying potential candidates for longevity trials, streamlined the monitoring of patient responses, and allowed for the integration of complex health data from various sources. In several trials, AI-driven tools have been credited with optimizing the outcomes by predicting adverse effects and adjusting trial protocols in real time (Esteva et al., 2017).

While these stories highlight success, they also underline the limitations inherent in current research. Longevity science in human trials must navigate the complexities of human biology—more convoluted than any model organism or theoretical calculation. Successes thus far have opened doors but also revealed new corridors of unknown, inviting researchers to walk through them with both excitement and caution.

The continuous evolution of these trials is essential as they offer a roadmap to future exploration. Each success story is a snapshot of where science stands in its timeless quest against aging. As society wrestles with these new realities, the importance of clear ethical frameworks and robust legal standards becomes evident. Only through thoughtful consideration can the potential of these human trials be fully realized while safeguarding against unintended societal consequences.

Despite these challenges, the progress seen in human trials teaches us that the pursuit of longevity is as much a story of human resilience and curiosity as it is an academic venture. The potential to alter the natural trajectory of aging is within sight, provided that society as a whole can come together to address the fundamental questions that arise from these advancements. Balancing scientific exuberance with philosophical reflection will ensure that the success stories of today lay a solid foundation for the breakthroughs of tomorrow.

It is clear that longevity science stands poised at the edge of what was once thought impossible. As we continue to glean insights from human trials, the horizon of what we can achieve—and understand—expands. The ultimate aim is not just to extend life but to improve the quality of those extra years. Building upon today's achievements, we can envision a future where aging is not seen as inevitable decline but a manageable phase of life, filled with opportunity and vitality.

Lessons from Animal Research

In our journey through the landscape of longevity science, there's much to learn from the animals that roam alongside us on this planet. These creatures often exhibit traits and lifespans that captivate the curiosity of scientists aiming to extend human lives. From jellyfish that seem to defy death to rodents with remarkable resilience, the animal kingdom offers a wealth of insights that can inform our understanding of human longevity.

One of the most astounding examples comes from the hydra, a diminutive freshwater polyp. Some species of hydra have exhibited negligible senescence, appearing not to age at all over their lifetimes (Martínez, 1998). This biological fountain of youth is linked to their ability to constantly regenerate their tissues, a feat enabled by an abundance of stem cells. Unlike humans, whose stem cell numbers dwindle with age, the hydra maintains its regenerative capability throughout its life. This phenomenon invites questions: Can humanity harness similar regenerative processes? Could our scientific expedition someday allow us to preserve our own cellular youthfulness?

Moreover, the intriguing case of the naked mole rat offers valuable lessons. Unlike their rodent relatives, naked mole rats enjoy significantly extended lifespans, with some living over thirty years. Researchers attribute this longevity to a mutation in a gene responsible for the production of high-molecular-weight hyaluronan, which seems to play a role in cancer resistance (Tian et al., 2013). This discovery has prompted investigations into the potential applications of hyaluronan in human medicine. We might ask, could this compound become a key ingredient in our quest to prevent cancer and other age-related diseases?

Another remarkable example is the longevity of certain species of tortoises and sea turtles. These reptiles can live for over a century, with their extended lifespans partly credited to their slow metabolisms and exceptional DNA repair mechanisms (Gavilán et al., 2018). By studying these traits, scientists are even uncovering mechanisms that might one day help in mitigating the human body's own decay over time. The question arises: how can we mimic such efficient DNA repair processes to stave off the ravages of aging?

Turning our attention to the depths of the ocean, we find the bowhead whale, an impressive mammal known for its lifespan exceeding 200 years. The genetic adaptations that enable their longevity and resistance to diseases are of immense interest. For instance, these whales possess genetic variations in pathways related to DNA repair, cell cycle regulation, and cancer suppression (Keane et al., 2015). Analyzing these genetic characteristics may open doors to new therapeutic targets for extending human life and delaying age-related conditions.

Though the field of animal research in longevity has its breakthroughs, it's not without its ethical quandaries. As we draw closer to understanding the secrets behind these creatures' extended lifespans, we must tread carefully. There's a delicate balance between scientific pursuit and moral responsibility. When exploring interventional possibilities in humans, what ethical ramifications must we consider? How do we ensure that our quest for longevity respects the wellbeing of all living creatures involved?

Furthermore, the cross-species comparisons necessitate caution. While many animals exhibit extraordinary longevity traits, translating these findings to humans is complex. Humans have different physiological and genetic makeups, and not all animal-based insights may be applicable to human biology. There's a need for rigorous validation and adaptation of these findings to ensure safety and efficacy in potential human applications.

With these varied lessons from the animal kingdom, a philosophical discourse beckons. There's a continuous tug-of-war between accepting our natural lifespans or pushing the boundaries of life itself. Our expedition into animal longevity illustrates a universe humming with possibilities – and challenges us to ponder the extent to which we wish to manipulate nature's timeline. Do we seek longevity for its own sake, or are we driven by deeper motivations to achieve something within our extended years?

Ultimately, while animals like the hydra, naked mole rat, and bowhead whale provide tantalizing clues, we must strive to integrate these lessons into a coherent strategy for human longevity. Advances in genetics, regenerative medicine, and biochemistry fueled by animal research are not the final destination but a part of the grand blueprint for human life extension.

So, as we delve into this formidable expedition with wide-eyed wonder, we should remain anchored in scientific rigor and ethical clarity. The insights drawn from animal research contribute significantly to the canvas of longevity science, each stroke painting a picture of what might yet be possible for humanity. But it's a picture that should always respect the natural order, where we stand not above it, but in conscious communion with the myriad forms of life that inspire our quest for enduring vitality.

Chapter 15:
Future Directions

As we stand on the precipice of remarkable scientific advancements, the future of longevity science is poised for transformative change. Emerging technologies such as advanced biotechnologies, artificial intelligence, and novel pharmacological interventions will likely redefine our understanding of aging and life extension (Johnson et al., 2023). These innovations promise a future where personalized medicine and genome editing could potentially render aging-related diseases obsolete, thereby extending the healthspan and lifespan of individuals (Smith & Chen, 2022). As we anticipate the next frontier, it's crucial to focus on integrating ethical considerations with technological prowess; while the promise of prolonged life is tantalizing, the implications of such advancements necessitate a balanced approach involving equitable access and societal readiness (Lee, 2021). The synergy of cross-disciplinary research and international collaboration will, no doubt, spearhead the evolution of this field, making the impossible not only imaginable but achievable. Yet, these aspirations must be tempered with caution, ensuring that the pursuit of longevity harmonizes with the broader narrative of human experience and values.

Emerging Technologies

As we stand at the precipice of unprecedented technological advancement, the realm of emerging technologies unfolds with both promise and complexity. The fusion of cutting-edge innovation with the quest for longevity may well redefine what it means to age and, indeed, to live. This section examines the frontier of such technologies and their potential applications in extending human life. Over the past few decades, breakthroughs in technology have gradually paved the road to what many now dream of—longevity not just in terms of years, but in quality of life as well.

One cannot discuss emerging technologies in longevity without acknowledging the rapid advancements in nanotechnology. Nanotechnology, with its ability to manipulate matter at the atomic level, offers significant potential in medicine and, by extension, the quest for immortality. By designing nanoparticles to target and repair damaged cells, we can foresee a future where age-related deterioration is not just slowed, but actively repaired. This precision, akin to a microscopic surgeon, opens up possibilities ranging from the reversal of cellular aging to the delivery of therapeutics that can stave off diseases previously deemed inevitable (Freitas, 2010).

Alongside nanotechnology are developments in bioinformatics and computational biology. Driven by the exponential growth of computational power and data analytics, bioinformatics has allowed scientists to analyze complex biological data at a scale never before achievable. This field empowers us to predict and model the biological mechanisms underpinning aging processes. With computational models, researchers can simulate how interventions at genetic or cellular levels might alter aging pathways, thus fast-tracking the development of new therapies without the years of clinical trials traditionally required (Kirk, 2011).

Biotechnology, especially in the realm of synthetic biology, is another frontier that promises significant advances in lifespan extension. By designing and engineering new biological parts and systems, synthetic biologists aim to create organisms with beneficial characteristics or improve existing organisms to enhance healthspan. Imagine a world where biological machines, designed by humans, repair the wear and tear of aging cells, maintaining optimal functionality throughout an individual's life. Such innovations could dramatically alter how we perceive and experience aging (Church & Regis, 2014).

At the intersection of AI and healthcare technology, we witness the emergence of personalized medicine as a transformative force in longevity. Artificial Intelligence, when integrated with massive datasets from health records, genetic information, and lifestyle data, offers deep insights into personalized risk factors and treatment responses. Machine learning algorithms can identify patterns and predict health outcomes, enabling personalized interventions that are more effective than the one-size-fits-all approach that dominates today's healthcare paradigm (Topol, 2019).

Robotics and automation also stand critical in this narrative. From robotic surgeons performing intricate procedures with unparalleled precision to exoskeletons offering mobility to the aging, robotics is revolutionizing the landscape of healthcare and aging. Automation in caregiving, through the deployment of AI-driven companions, promises to alleviate some of the burdens associated with elder care, providing not only physical assistance but also emotional and cognitive support (Cohen, 2020).

The prospect of quantum computing, though in its infancy, offers an exciting potential for revolutionary breakthroughs in longevity science. By exponentially increasing computational power, quantum computing could solve complex biological puzzles at speeds unimaginable today, such as protein folding or complex simulations of drug interactions. These systems could unravel the codes that control aging processes, providing insights and solutions not currently accessible (Preskill, 2018).

Furthermore, the rise of blockchain technology offers an innovative approach to managing health data in a secure, transparent, and decentralized manner. With patient data stored and shared transparently, research in aging can proceed with enhanced data integrity and collaboration across global institutions, potentially accelerating the pace of discoveries in age-related treatments (Kuo et al., 2017).

Despite these promising technologies, we must navigate the ethical landscape with caution. The convergence of these fields raises significant questions about access, equity, and the unintended consequences of dramatically extended human lifespans. As technology races forward, society must deliberate on how these advancements are regulated, ensuring they serve the greater good without exacerbating existing inequalities (Kass, 2016).

In contemplating the potential of emerging technologies, we approach a thrilling yet daunting frontier. The interplay of scientific discovery, ethical considerations, and philosophical insights will shape how humanity navigates this brave new world. These technologies collectively offer a vision not just of extended life, but of life enhanced in quality and fulfillment—an aspiration that stretches beyond the confines of time and into the realm of possibility.

The journey of integrating these technologies into the fabric of society and individual lives remains complex and filled with challenges. Yet, it's in this complexity that we find the impetus to strive forward. The emergence of these technologies compels us to reflect on not only the future of longevity but what it fundamentally means to be human in an era defined by rapid technological change and boundless potential.

The Next Frontier in Longevity Science

As we stand on the precipice of unprecedented scientific advancements, the quest to extend human life reveals ever more complex challenges and opportunities. The promise of longevity is not solely a pursuit of years, but an endeavor to add vitality and quality to those years. Scientists and researchers across the globe are relentlessly exploring the uncharted territories of human biology in hopes of unlocking the secrets of prolonged and healthier lives. This journey into the unknown is the next frontier in longevity science, a field teetering on the edge of breakthrough innovations and existential questions.

The combustion of scientific curiosity fuels the journey into the future of longevity, with biology and technology working symbiotically. One major area of exploration involves unraveling the intricate mechanisms of aging at a molecular level. Aging, once thought an irreversible process, is being dissected down to its very genes and cellular processes. Here, the convergence of disciplines like genetics, biochemistry, and computational science signify the potential for interventions that could edit life's script at its most fundamental levels (Lopez-Otin et al., 2013).

Telomere biology illustrates the immense potential within our own DNA, shedding light on cellular aging and mortality. Telomeres, the protective caps at the end of chromosomes, have been likened to the plastic tips of shoelaces, preventing chromosomal fraying and fusion. With every cell division, these telomeres shorten, eventually triggering senescence or programmed cell death. Researchers investigate ways to preserve telomere length and enhance telomerase activity, aiming to fend off the effects of aging (Harley, 1991).

Similarly, the field of epigenetics offers promising prospects in reprogramming the aging process. Unlike genetic mutations, epigenetic changes do not alter DNA sequences but modify how genes are expressed. Environmental and lifestyle factors can influence these changes, thus it becomes conceivable to modulate them to delay or reverse aging. Techniques like CRISPR are being adapted to edit not just genes, but the epigenetic markers themselves, offering unprecedented precision and hope (Simmons, 2008).

The exploration extends outward from our genes to our gut, as emerging research underscores the gut microbiome's profound impact on longevity. A thriving, biodiverse gut microbiome contributes to robust immune function, efficient digestion, and even mental health. Scientists are delving into how manipulating the gut microbiome through diet, probiotics, and prebiotics, can extend life and enhance overall well-being (O'Toole & Jeffery, 2015).

In addition to biological inquiries, technological advancements such as the deployment of artificial intelligence (AI) in longevity research mark a pivotal advancement. AI-driven models refine predictions about aging processes and enhance the accuracy of personalized medicine. By integrating vast repositories of biomedical data, AI systems can predict disease onset, outcomes, and treatment responses, potentially identifying novel targets for longevity therapeutics (Topol, 2019).

The integration of nanotechnology in medicine also hails potential transformative impacts on longevity. Nanoparticles could deliver therapies with precision at the cellular level, repairing tissues and eradicating senescent cells that contribute to aging and chronic disease. Innovations in nanomedicine offer the tantalizing prospect of targeted interventions that could delay aging and enhance human healthspan (Riehemann et al., 2009).

As we venture deeper into this new frontier, the question of how these advancements will translate across diverse populations presents both an opportunity and a challenge. While scientific and technological breakthroughs offer potential universal benefits, equitable access to these treatments is imperative. Ensuring accessibility and affordability requires global collaboration between scientists, governments, and industry leaders. Only by doing so can we ensure that longevity science contributes to reducing health disparities rather than exacerbating them.

The dialogue surrounding the ethical implications of life extension is equally critical. As science rewrites the boundaries of life's possibilities, philosophical and moral frameworks must evolve concurrently. What would it mean to extend life significantly? How would it transform our culture, familial structures, and economic systems? As researchers forge ahead, these questions necessitate wide-ranging discussions that encompass both the societal implications and the individual aspirations for a longer life.

Wherever the voyage into the future of longevity science leads, one constant remains: the human spirit's tenacity and creativity. Meeting the challenges of this next frontier requires not just scientific acumen, but a collective vision of what it means to live well and long. It demands that we tread a path that respects both our deepest ambitions and our shared humanity. This journey, uniting the fundamental sciences and ethical discourse, promises not just to extend life, but to enrich it in ways yet unimagined.

Chapter 16:
Risk Management in Longevity

In the pursuit of extending human lifespan, the intricacies of risk management become an essential component, bridging science and philosophy in a dance of proactive caution. As we stand on the precipice of unprecedented technological advancements, from genetic modification to AI-driven health solutions, the potential for both remarkable progress and unforeseen complications demands a profound ethical oversight. This balance is crucial to navigate the labyrinth of potential risks associated with longevity science, ranging from genetic anomalies to societal inequalities (Smith et al., 2020). To mitigate such risks, multifaceted strategies are integral, including cross-disciplinary collaborations that encompass not only scientists and technologists but also ethicists and policymakers. Only through a comprehensive approach can we align the objectives of prolonged human life with moral and societal principles, ensuring that the treasure of extended lifespan does not become a Pandora's box but remains a beacon of hope for future generations.

Ethical Oversight

The integration of ethical oversight in longevity science is not just a moral obligation but a strategic necessity. As researchers delve into the secrets of extending life, it's imperative to scrutinize the ethical dimensions that accompany such undertakings. Longevity science sits at the nexus of biology, technology, and philosophy, beckoning us to consider not only what we can do but what we should do. This chapter will explore the diverse strata of ethical oversight necessary to navigate these uncharted territories.

Above all, ethical oversight in longevity science must address issues of accessibility and inequality. Achievements in extending human life should not be restricted to the affluent, creating a gap similar to the digital divide but more severe. The advancements in longevity treatments risk becoming exclusive commodities, available only to the wealthy, thus widening societal and health disparities. Ethical oversight must ensure that new technologies do not exacerbate existing inequalities but rather aim to reduce them. Researchers and policymakers alike must consider frameworks that guarantee fair access (Juengst, 2020).

Furthermore, ethical oversight must engage with the potential societal impacts of increased human lifespan. While the idea of living longer is enticing, it's consequential; people could face challenges relating to economic systems, job markets, and societal structures. A longer lifespan may lead to extended periods of work, delayed retirements, and shifting family dynamics. Ethical frameworks ought to address these issues head-on, integrating them into policy recommendations that account for societal evolution over time.

In addition, within the realm of ethical oversight lies the question of consent. With advancements such as genetic manipulation and AI-driven health monitoring, obtaining informed consent becomes notably complex. It's crucial that individuals truly understand the potential ramifications of longevity treatments. The notion of informed consent should evolve alongside technological advances to encapsulate a complete understanding of benefits, risks, and unknowns tied to these emerging technologies (Vayena et al., 2018).

Ethical oversight also involves the environmental implications of a significantly expanded human lifespan. With more people living longer lives, the strain on natural resources and ecological systems could intensify. What constitutes a sustainable life extension? Ethical considerations should include the long-term impact on the planet, urging balanced approaches that weigh human longevity against environmental sustainability. This dimension emphasizes the interconnectedness of our biological persistence and ecological stewardship (Kass, 2003).

Moreover, the ethical landscape in longevity science must contemplate the philosophical underpinnings of advancing human life. The very pursuit raises existential questions about the meaning and value of life—questions that have occupied philosophers for millennia. Ethical oversight should not just be about managing risks, but also about guiding scientific endeavors in alignment with broader human values. This calls for a multidisciplinary approach, engaging philosophers, ethicists, and theologians alongside scientists.

Ethical oversight committees in longevity research must be composed of diverse stakeholders to ensure a multi-faceted perspective. Scientists, ethicists, sociologists, and even representatives from the public should form part of these committees, promoting transparency and accountability. A well-rounded composition of oversight bodies can help bridge the gap between scientific possibilities and ethical responsibilities, fostering public trust in longevity science (Juengst & Benjamin, 2014).

Public engagement is another pillar of ethical oversight. As longevity science potentially reshapes human experience, involving the public in dialogues about its significance is key. Public forums, educational programs, and open-access publications can demystify the science, making it accessible and understandable to laypeople. An informed public is more likely to contribute positively to discussions and policies surrounding the ethical use of longevity technologies.

Lastly, ethical oversight must adapt to the changing landscape of longevity research. As new discoveries unfold, ethical considerations should not be static; they must be dynamic, evolving alongside scientific breakthroughs. Regular revision and incorporation of new ethical guidelines will ensure that the field not only stays scientifically robust but also ethically sound.

In conclusion, ethical oversight in longevity science must be comprehensive, proactive, and inclusive. By addressing issues of access, societal impact, consent, environmental sustainability, philosophical implications, stakeholder diversity, public engagement, and adaptability, we can strive for a future where the extension of human life is not just a scientific triumph but an ethical achievement that benefits all.

Mitigating Technological Risks

In the intricate world of longevity science, technological advancements promise remarkable opportunities to extend human life. However, these innovations also present challenges that need careful management. A strategic approach to understanding, assessing, and mitigating technological risks can ensure that the quest for extending human life does not derail due to unforeseen pitfalls.

The rapid pace of technological innovation brings a dual-edged sword. On one hand, it fuels our ambition to prolong life, but on the other, it creates new vulnerabilities. Consider the development of genetic interventions like CRISPR, which holds the potential to eliminate genetic diseases and decelerate aging. However, without meticulous regulation and ethical oversight, the misuse of such technologies could lead to unforeseen consequences, like genetic inequality or ecological imbalance (Lanphier et al., 2015).

Establishing comprehensive regulatory frameworks is crucial in mitigating these risks. These frameworks must not merely stifle innovation; instead, they should provide guidelines that ensure safety and equity in technological applications. The challenge is to develop regulations that can keep pace with the speed of scientific advancements, ensuring that as new technologies emerge, they are thoroughly evaluated for potential risks and benefits (Collins, 2020).

Another crucial aspect is the secure management of data generated through these technologies. With AI-driven health monitoring systems becoming more prevalent, vast amounts of personal health data are collected. Ensuring the privacy and security of this data is paramount. Data breaches or misuse could lead to significant personal and societal harm, raising ethical questions about surveillance and consent (Li et al., 2021).

Moreover, fostering interdisciplinary collaboration is key to building robust risk management strategies. When technologists, ethicists, policymakers, and the public engage collectively, the development of socially responsible innovations becomes feasible. Collaborative dialogue not only enhances the understanding of potential risks but also encourages the development of creative solutions to mitigate them.

Furthermore, public education plays a foundational role in this process. By demystifying the technologies involved in longevity science, the public can make informed decisions and actively participate in legislative processes. Indeed, a well-informed public acts as a safeguard against hasty implementation of high-risk technologies and helps build a supportive environment for responsible technological experimentation.

Technological risk mitigation in the context of longevity also involves preparing for the unintended consequences that may arise once a technology becomes widespread. Scenario planning and risk assessment models can be employed to foresee potential outcomes and prepare contingency plans accordingly. This proactive approach not only mitigates risks but also enhances the resilience of society in adapting to new technologies.

In balancing innovation with caution, it becomes essential to foster a culture of ethical accountability among scientists and technologists. Academic institutions and research organizations could develop ethical training programs, emphasizing the larger societal implications of technological use. This could cultivate a breed of professionals who are not only technologically proficient but also ethically conscious, steering the development of safe and equitable longevity technologies.

Additionally, international cooperation is vital in handling technological risks associated with longevity. Since scientific progress knows no national boundaries, the risks and benefits are globally shared. International treaties and agreements can provide a platform for consensus-building on safety standards and ethical considerations, ensuring that the global community moves forward together in the advancement of safe life-extending technologies.

In conclusion, while the allure of significantly extending human lifespan is compelling, it is crucial to address the inherent technological risks with diligence and foresight. By establishing robust regulatory frameworks, securing data privacy, fostering interdisciplinary collaboration, educating the public, and promoting ethical accountability, we can navigate the complex landscape of longevity technologies safely and responsibly. These efforts will not only safeguard against potential risks but also pave the way for a more equitable and sustainable future where the promise of longevity can be realized without undermining societal well-being.

Chapter 17:
Aging and Society

As we delve into the complex interplay between aging and society, it's essential to acknowledge the profound impact of a rapidly aging global population on social structures, economies, and cultural norms. This demographic shift presents both an aging population crisis and opportunities for sustainable solutions. The increase in life expectancy has inevitably led to a strain on healthcare systems, pensions, and labor markets, urging us to reconsider how we support older generations (Bloom et al., 2011). However, it's not merely a challenge but a call to innovate and adapt our societal frameworks. We must foster environments where aging individuals can maintain their health, independence, and dignity, contributing to society in meaningful ways. By integrating technology with compassionate human-centered policies, societies can transform this potential crisis into an era of enriched human experience, provided they navigate these changes thoughtfully (Fukuyama, 2002). The intersection of aging within societies offers a unique lens to rethink the longevity discourse, underscoring the necessity for comprehensive solutions that ensure equity and sustainability for future generations.

The Aging Population Crisis

The growing global population of older individuals presents a profound challenge that societies across the globe must contend with. In recent decades, advancements in healthcare and an increased standard of living have led to remarkable increases in life expectancy. While this is a testament to human progress, it also adds layers of complexity to existing socio-economic fabrics. As the median age rises, the implications for economies, healthcare systems, and social structures become more pressing.

One of the most immediate concerns is the economic impact of an aging population. An increased proportion of elderly individuals means a higher dependency ratio, with fewer working-age people supporting a larger retired population. Countries with significant social security systems face mounting pressure to maintain their commitments. In the United States, for instance, the Social Security Administration has long warned that its trust funds may be depleted in the near future if no corrective measures are taken (Social Security Administration, 2021). This phenomenon isn't unique to the U.S.; nations worldwide are grappling with similar fiscal stress.

The economic challenges don't end there. As individuals age, they invariably face rising healthcare needs. Chronic conditions, more prevalent among older adults, contribute heavily to increased healthcare expenditure. A study by Thorpe and Howard (2006) showed that a significant portion of the healthcare cost surge is attributable to chronic illnesses prevalent in aging adults. This places an unprecedented burden on healthcare infrastructures, which must adapt to increased demand while innovating to provide cost-effective care solutions.

Beyond economics and healthcare, the social ramifications of an aging population are multifaceted. The traditional family structure undergoes a transformation as families cater to the needs of older members. The "sandwich generation" finds itself supporting both aging parents and their own children, leading to emotional and financial strain. On a societal level, there is a risk of intergenerational tensions as resource allocations become a matter of public debate (Harwood, R. H., 2019).

Furthermore, the workforce dynamics shift significantly. As experienced employees retire, the transfer of knowledge and skills becomes a critical business concern. Countries with aging populations must also confront the potential skill shortages that arise from a dwindling younger workforce. Some nations, notably Japan and Germany, have initiated policies to encourage workforce participation among older adults, including flexible work arrangements and incentives for delayed retirement (Niehues & Strube, 2019).

Yet, within these challenges lie opportunities for innovation and societal growth. For example, the rise of the "silver economy" highlights the market potential driven by seniors. Their significant purchasing power catalyzes industries such as healthcare, real estate, and tech. Products and services catering specifically to the needs and preferences of the elderly population are burgeoning, from smart home technologies designed for accessibility to travel and leisure offerings targeting seniors.

From a philosophical perspective, the aging population invites us to reassess our values and societal priorities. It challenges the prevailing notion of productivity, proposing that aging can be a time of reflection, wisdom, and mentorship. Respect and integration of elderly individuals into society can enhance communal ties, with countries like Sweden illustrating the benefits of intergenerational interaction in strengthening social cohesion (Hofäcker, 2010).

Moving forward, sustainable solutions for the aging population crisis must consider holistic approaches. Policies should aim not just to manage the growing needs but also to empower older adults to live fulfilling lives. This includes promoting active aging and lifelong learning, ensuring seniors retain a sense of purpose and contribution. Public and private entities must collaborate in innovative caregiving models, potentially leveraging technology for remote health monitoring and care coordination.

In addressing the aging population crisis, international collaboration becomes indispensable. Countries must share insights and strategies, learning from successful models while adapting them to local contexts. The global nature of the challenge requires a sharing of resources and knowledge, transitioning from insular national strategies to cooperative international efforts.

Ultimately, the discussion surrounding the aging population crisis is one of adaptation and resilience. As societies evolve to accommodate a new demographic reality, the potential exists not only to mitigate challenges but to embrace the benefits an aging population can bring about. The journey forwards demands creativity and compassion to redefine what it means to grow old in the twenty-first century.

Solutions for Sustainable Aging

As the global population ages, the challenge of ensuring sustainable aging has become increasingly pressing. Sustainable aging is not just a health concern; it's a multidimensional issue that touches on economic, social, and environmental domains. Addressing this requires integrated solutions that not only aim for longevity but also enhance the quality of life for the elderly. As we embark on this journey towards sustainable aging, it's essential to explore avenues that align with societal capabilities and resource constraints to ensure the future is both equitable and manageable.

One promising solution for sustainable aging lies in the development of age-friendly communities. These are not merely places where seniors live; they are environments designed specifically to accommodate the needs of older adults while fostering their engagement in community life. By prioritizing accessibility and inclusivity in urban planning, we can create spaces that encourage active living and social interaction, thereby combating the isolation that often accompanies aging. Initiatives that improve public transportation, offer mixed-use developments, and promote walkable neighborhoods can significantly enhance the daily lives of aging individuals (World Health Organization, 2007).

Healthcare innovation is another vital component of sustainable aging. Advances in telemedicine and personalized medicine hold significant promise for revolutionizing how we care for the elderly. Remote healthcare solutions, equipped with AI-driven diagnostics and wearable technology, can facilitate continuous monitoring of health conditions without the need for frequent hospital visits. This ensures timely intervention and continuity of care, reducing the burden on healthcare systems while improving patient outcomes (Topol, 2015). Personalized medicine, focusing on tailored treatments based on genetic and lifestyle factors, also allows for more efficacious management of age-related diseases.

The economic aspect of sustainable aging can't be overlooked, and it requires careful strategy. Aging populations may impose a strain on public finances, particularly in healthcare and pension systems. To counteract this, innovative funding models and policy frameworks are essential. This includes pension reforms that reflect increased life expectancy and the promotion of lifelong learning and employment opportunities for older adults, which can keep them engaged in the workforce longer. Not only can this alleviate some financial pressures, but it also empowers seniors to contribute meaningfully to society (Bloom et al., 2011).

Social support systems are equally critical in the quest for sustainable aging. Intergenerational programs that bridge the gap between youth and the elderly can foster mutual understanding and support. Programs that encourage shared housing, for instance, benefit both younger and older generations, providing affordable housing solutions while reducing loneliness and building community ties. Moreover, such initiatives promote cultural exchanges and the transmission of knowledge and values (Hagestad & Uhlenberg, 2005).

Environmental sustainability also plays a remarkable role in sustainable aging. Climate change and environmental degradation disproportionately affect the elderly, who are more vulnerable to extreme weather events. Incorporating green spaces in urban planning not only caters to physical well-being through opportunities for recreation and exercise but also supports mental health, as nature exposure is known to reduce stress and promote well-being. Sustainable energy solutions and eco-friendly housing designs can also ensure that living spaces remain comfortable and affordable, even as energy costs and climate conditions evolve (Pretty et al., 2007).

Technological literacy and access for the elderly remain a backbone of this conversation. Education programs and community centers dedicated to teaching seniors how to use emerging technologies can bridge the digital divide. This empowerment enables older adults to more easily communicate with family and friends, access services, and remain informed about health and wellness. Technology can facilitate lifelong learning, ensuring that the elderly remain intellectually engaged and socially connected, thus enjoying a more fulfilling life.

A philosophical perspective on sustainable aging invites us to rethink aging not as a decline but as a new chapter for potential growth and contribution. By adopting a cultural shift in how we perceive aging, the stigma associated with growing older can diminish. Celebrating aging as a natural progression and harnessing the wisdom and experience of the elderly provides a unique opportunity to enrich society. Encouraging narratives that paint aging in a positive light can transform public perceptions and attitudes, nurturing a society that truly values all its members.

This holistic approach to sustainable aging demands cross-sector collaboration and innovation. Policymakers, healthcare providers, urban planners, and community leaders must work in concert to craft multi-faceted solutions that address the complex needs of an aging society. This not only involves addressing the obvious physical and health-related challenges but also the social, economic, and environmental aspects of aging. Only through such a comprehensive approach can sustainable aging move from ideal to reality.

In conclusion, sustainable aging is possible through a tapestry of interconnected solutions tailored to address the nuanced needs of the aging population. By cultivating age-friendly communities, leveraging healthcare innovations, implementing strategic economic policies, fostering social support, ensuring environmental adaptability, and embracing technology, we can create a future that honors and sustains the dignity and vitality of the aging. As society continues to embrace these challenges, the potential to thrive in later years becomes not just an individual triumph but a collective achievement.

Chapter 18:
Innovations in Healthcare

The landscape of healthcare is undergoing a transformative shift as innovations redefine how we understand and manage human health. At the forefront of this evolution is personalized medicine, where treatments are tailored to individual genetic profiles, allowing for more precise and effective interventions. This approach promises to revolutionize the traditional one-size-fits-all treatment model, increasing efficacy while minimizing side effects (Katsios, 2020). Concurrently, the advent of remote health solutions leverages technology to provide access to care almost ubiquitously, dismantling geographical barriers and facilitating continuous patient monitoring. As telemedicine, wearable health devices, and AI-driven diagnostics become more sophisticated, they not only empower patients with information but also enhance the capacity of healthcare systems to deliver timely and personalized care (Smith & Lee, 2021). While these innovations hold immense promise, they also demand rigorous oversight to balance technological advancement with ethical considerations. This dynamic interplay between innovation and regulation will shape the future of healthcare, steering humanity toward a new era of health and well-being.

Personalized Medicine

Personalized medicine is at the forefront of contemporary healthcare innovations, promising to revolutionize patient care by shifting from a one-size-fits-all approach to one that is as unique as each individual. This approach integrates genetic, environmental, and lifestyle factors into medical decision-making, offering a tailored fit for prevention, diagnosis, and treatment ("Collins & Varmus, 2015"). As technology advances, the potential for personalized medicine to increase efficacy while minimizing adverse effects is becoming ever more tangible.

The foundation of personalized medicine lies in understanding the genetic makeup of individuals. With the rapid advancement of genomic technologies, such as next-generation sequencing, the cost of sequencing an individual's genome has plummeted. This makes it practical for widespread clinical application. It allows for the identification of genetic variations that can predict the likelihood of developing certain diseases, as well as the patient's likely response to specific medications ("Mardis, 2018"). This knowledge empowers healthcare professionals to design more effective and customized treatment plans.

One exciting implication of personalized medicine is pharmacogenomics—the study of how genes affect an individual's response to drugs. By leveraging this understanding, clinicians can prescribe medications that are best suited to an individual's genetic profile. This could lead to reduced trial-and-error prescribing, a common practice in traditional medicine, and minimize the risk of adverse drug reactions ("Spear et al., 2001"). For instance, individuals with certain genetic profiles might metabolize drugs at different rates, influencing both efficacy and safety. These insights can transform how conditions like cancer, cardiovascular diseases, and mental health disorders are managed.

Beyond genomics, personalized medicine integrates environmental and lifestyle factors—two key determinants of health. A person's environment and lifestyle choices, such as diet, exercise, and exposure to toxins, interact with their genetic predispositions, influencing disease risk and progression. By utilizing data from wearable technology and other monitoring devices, healthcare providers can obtain a holistic view of these influences, leading to more precise health recommendations ("Topol, 2019").

The application of personalized medicine in oncology is perhaps the most notable example of its transformative potential. Cancer, characterized by genetic mutations that fuel uncontrolled cell growth, naturally lends itself to this customized approach. Targeted therapies, developed through the understanding of unique genetic markers present in different cancer types, have shown notable success. Drugs like trastuzumab and imatinib stand as testaments to the efficacy of treatments aligned with genetic diagnostics ("Sawyers, 2004"). However, the journey is only beginning, and with ongoing research, these successes pave the path for more breakthroughs.

Personalized medicine brings with it philosophical and ethical considerations, particularly around data privacy and access. The extensive use of genetic data necessitates robust privacy measures to prevent unauthorized access and potential misuse. It also raises questions of equity: who gets access to these cutting-edge medical advancements? Efforts must be made to ensure that personalized medicine doesn't exacerbate existing healthcare disparities but instead becomes a universal standard ("Knoppers, 2014"). Policies and frameworks need to be devised to manage these issues effectively and equitably.

Another ethical dimension relates to the psychological implications of genetic information. How patients perceive and handle information about their genetic predispositions can significantly impact their mental well-being. Predictive knowledge of potential diseases can be empowering but also burdensome. It requires a sensitive and supportive approach from healthcare providers to guide patients through understanding and action ("McBride et al., 2010").

This highlights the need for more personalized communication and care strategies within the realm of personalized medicine.

Education and collaboration are critical in advancing personalized medicine. Multidisciplinary approaches, integrating researchers, clinicians, and data scientists, facilitate the translation of scientific discoveries into practical medical applications. Furthermore, educating healthcare professionals—ensuring they are well-versed with the principles and tools of personalized medicine—is essential for its successful implementation in daily practice ("Flores et al., 2013").

Looking forward, the integration of artificial intelligence and machine learning into personalized medicine promises even greater strides. These technologies can process and analyze vast datasets far more efficiently than human capability alone, allowing for deeper insights into genetic patterns and disease mechanisms ("Obermeyer & Emanuel, 2016"). The symbiosis between AI and personalized medicine can refine the precision of diagnostics and treatments, pushing the boundaries of what's possible in patient care.

In conclusion, personalized medicine heralds a transformative era in healthcare, combining the intricacies of an individual's genetic blueprint with their environmental contexts and personal habits to craft bespoke medical solutions. The journey from conceptual framework to ubiquitous application is fraught with challenges, yet filled with promise. The ability to not only understand the genomic puzzle pieces but know how they fit within the broader tapestry of health and wellness, speaks to a future where medicine genuinely mirrors the individuality of human life.

Remote Health Solutions

As we venture further into the 21st century, the integration of technology in healthcare is transforming not only how we approach medical treatment but our very definition of wellness. Remote health solutions have emerged as a pivotal innovation, offering the potential to bridge gaps in access, provide real-time monitoring, and personalize treatments in ways previously unimaginable. These innovations embody a shift from traditional, reactive healthcare to a more proactive, patient-centric model.

Modern society faces a myriad of challenges that necessitate remote health solutions. Chronic diseases, aging populations, and geographical barriers are just a few of the major hurdles. With these challenges come opportunities for technology to enhance patient care. Telemedicine, for instance, has been a revolutionary force, allowing patients to consult healthcare professionals without the constraints of location (Bashshur et al., 2016). This development represents a democratization of healthcare, making it accessible in both urban centers and remote rural areas.

Remote monitoring technologies are equally transformative. Devices that monitor vital signs and transmit data in real-time empower patients with chronic conditions such as diabetes or hypertension. These tools enable continuous tracking and early interventions, which can prevent complications and reduce hospital admissions (Piette et al., 2015). The implications for improved patient outcomes and reduced healthcare costs are profound.

Moreover, remote health solutions align with the principles of personalized medicine, a concept gaining traction in the medical community. By utilizing wearable devices and sensor technologies, healthcare providers can tailor treatment plans based on an individual's unique health data and lifestyle factors. This personalization not only enhances treatment efficacy but also strengthens the patient's role in managing their own health.

The philosophical implications of remote health solutions are significant as well. They challenge traditional notions of the doctor-patient relationship, prompting questions about authenticity, empathy, and the nature of care. Can digital interactions genuinely replicate face-to-face consultations? While some argue that technology can lead to a depersonalized healthcare experience, others contend that it merely redefines the relationship, offering new avenues for maintaining human touch through digital means.

Despite the promise of remote health solutions, they are not without challenges. Data privacy and security are paramount concerns, as the influx of personal health data means increased vulnerability to cyber threats. Additionally, there is a risk of exacerbating existing inequalities if these technologies are not universally accessible. Policymakers and industry leaders must work together to devise strategies that ensure equitable distribution and safeguard sensitive information (Raghupathi & Raghupathi, 2014).

Another consideration is the integration of Artificial Intelligence (AI) in remote health solutions. AI has the potential to analyze large volumes of health data, predict outcomes, and even suggest treatment options. Yet, this raises ethical issues regarding decision-making and autonomy. For instance, should AI solely determine a patient's treatment path, or should it merely act as an assistant to human practitioners?

To address these challenges, collaboration across sectors is essential. Tech companies, healthcare providers, and governments must converge to create robust frameworks that encourage innovation while protecting individual rights. Standardizing protocols, ensuring interoperability among devices, and investing in infrastructure are all critical steps in this direction.

In a broader context, remote health solutions may catalyze a cultural shift in how health and wellness are perceived. By empowering individuals with tools to monitor and manage their health proactively, society could move towards a model where health is seen as a continuous, everyday responsibility rather than an occasional concern. This shift could redefine value systems around health, encouraging preventive measures and holistic well-being.

Ultimately, the future of remote health solutions is intertwined with the evolving landscape of healthcare itself. As we harness the power of technology to enhance medical treatments, the philosophical, ethical, and practical dimensions of these changes must remain at the forefront of our considerations. It is only by navigating these complexities thoughtfully that the full potential of these innovations can be realized to not just extend life, but to enrich its quality.

Chapter 19:
Nutrition and Longevity

In the intricate tapestry of human existence, nutrition stands as one of the pillars fundamentally shaping the journey of longevity. As we journey deeper into the exploration of life extension, it becomes increasingly apparent that what we consume plays a pivotal role in either accelerating or decelerating the aging process. A growing body of scientific literature suggests that specific diets can not only prevent chronic diseases but may also actively promote a longer lifespan (Fontana et al., 2014). The Mediterranean diet, rich in fruits, vegetables, whole grains, and healthy fats, is consistently associated with enhanced health outcomes and increased longevity. Likewise, studies on caloric restriction and intermittent fasting reveal pathways to longevity, hinting at their role in transcending mere survival to embrace thriving across an extended life (Longo & Panda, 2016). Furthermore, the rise of nutritional supplements, from antioxidants to omega-3 fatty acids, underscores a collective endeavor to harness scientific insights for prolonging vitality (Smith et al., 2020). As we unravel the complexities of nutrition, it becomes evident that our choices at the dining table are inextricably linked to the duration and quality of our lives, inviting us to consider food not merely as sustenance but as a formidable ally in the quest for longevity.

Diets that Promote Lifespan

As we continue our exploration of longevity, it's impossible to ignore the critical role of diet in promoting a long and healthy life. What we consume daily forms the bedrock of our physiological functioning and is deeply intertwined with our potential lifespan. The maxim "you are what you eat" isn't just a catchy phrase but a guiding principle backed by science. Our dietary choices influence our cellular health, genetic expression, and even our psychological well-being, underscoring the profound impact of nutrition.

One of the most talked-about dietary approaches that have gained popularity for its supposed longevity benefits is the Mediterranean diet. Originating from the eating habits of people in countries surrounding the Mediterranean Sea, this diet emphasizes a high intake of fruits, vegetables, whole grains, and healthy fats, particularly from olive oil. Moderate consumption of fish and poultry provides protein, while dairy and red meat are consumed sparingly. The Mediterranean diet is not just about specific food choices, but also about a lifestyle of slow, communal meals and mindful eating. Studies have linked this diet to decreased risks of heart disease, certain cancers, and neurodegenerative diseases (Martínez-González et al., 2017).

Similarly, the traditional diets of Okinawa, Japan, have been associated with exceptional longevity. The Okinawan diet is characterized by its low-caloric intake and high-nutrient density, primarily from sweet potatoes, green leafy vegetables, soy-based foods, and occasional fish. The limited use of sugar and processed foods contributes further to its health benefits. Intriguingly, the concept of "hara hachi bu," or eating until you are 80% full, is a social norm that has helped prevent overeating and maintains a healthy weight for the residents (Willcox et al., 2014).

Calorie restriction (CR) is another dietary intervention showing promise in extending lifespan, not just in humans but across multiple species. CR involves reducing calorie intake without malnutrition, thereby potentially slowing the metabolic rate and reducing oxidative damage in cells. While long-term studies in humans are ongoing, animal models, particularly rodents, have demonstrated that a significant reduction in calorie intake can extend lifespan by up to 40% (Fontana & Partridge, 2015). However, it's worth noting the challenges in adherence and the potential downsides of CR, such as decreased muscle mass and energy levels.

In addition to these well-known dietary strategies, emerging evidence suggests that intermittent fasting may offer significant longevity benefits. Intermittent fasting alternates between periods of eating and fasting, helping regulate insulin sensitivity and reduce inflammation. Although various fasting regimes exist, the concept remains consistent—providing the body a chance to restore and repair at a cellular level. Studies have shown that intermittent fasting can improve metabolic parameters and extend lifespan in animal models, although more research is needed in humans to fully understand its long-term effects (Anton et al., 2017).

Beyond specific dietary practices, the concept of plant-based diets has been gaining traction. Diets rich in fruits, vegetables, nuts, seeds, and legumes, while low in animal products, are associated with lower incidences of chronic diseases and, consequently, longer lifespans. The high fiber content and the presence of phytochemicals in plant-based foods support gut health and reduce systemic inflammation, contributing to the prevention of age-related diseases (Levine et al., 2014).

However, it's essential to consider that dietary habits' influence on longevity isn't just a matter of nutrient intake but also involves complex interactions with genetics and lifestyle. Personalizing dietary plans considering genetic dispositions and individual metabolic profiles can further optimize the benefits. Emerging fields like nutrigenomics explore how individual genetic variability can affect nutrient metabolism and influence health outcomes, paving the way for personalized dietary recommendations.

Moreover, the social and cultural contexts of eating cannot be overlooked. Eating is not just a biological necessity but a profound social experience. Social meals contribute not just to emotional well-being but also foster a balanced approach to eating, which can mitigate the risks of obesity and metabolic syndrome. Cultural practices and traditional dietary habits are woven into the fabric of community life, and these practices often reflect centuries of accumulated wisdom regarding health and diet.

Longevity diets also draw attention to the balance between quantity and quality. Consuming nutrient-dense foods rather than calorie-dense ones ensures that the body's nutritional needs are met without excess caloric intake. Monitoring and adjusting macronutrient ratios, such as reducing excessive sugars and consuming adequate, healthy fats, play into the broader strategy of aligning diet with longevity goals.

As science continues to reveal the intricacies of how our nutritional choices affect our lifespan, a harmonious connection between scientific understanding and practical application is vital. As people become more informed about the links between diet and longevity, there's a collective shift toward embracing diets that not only fuel longevity but also enhance the quality of life. The way forward involves continued research to refine these dietary approaches and adapt them to align with modern lifestyles, ensuring they remain sustainable and accessible to all.

Ultimately, embracing a diet that promotes lifespan is about making informed, mindful choices that consider both the scientific evidence and personal values. It's about fostering a relationship with food that goes beyond sustenance to embody a holistic strategy for long, vibrant living.

Nutritional Supplements

In the fascinating tapestry of human longevity, nutritional supplements have emerged as both a promising possibility and a subject of intense scrutiny. As the quest for prolonging life weaves through the corridors of science, philosophy, and advancement, nutritional supplements stand at the intersection of traditional wisdom and modern innovation. While they might not be the elixir of immortality that has mesmerized humanity through history, they offer a pathway to enhanced health and potentially increased lifespan, one rooted in the intimate relationship between nutrition and body functions.

The allure of nutritional supplements lies in their potential to fill gaps that our diet may not adequately cover. Contemporary lifestyles, often characterized by fast-paced routines and processed foods, have led to deficiencies in vital nutrients that are pivotal to maintaining cellular health. Supplements such as omega-3 fatty acids, vitamins D and B12, magnesium, and zinc are examples often touted for their health benefits (Fukuwatari et al., 2017). Each plays a unique role: omega-3s are crucial for cardiovascular and cognitive health, vitamin D aids in calcium absorption and bone health, while zinc supports immune function. The synergy of these supplements and their targeted applications could advance our understanding of extending healthy years.

Nonetheless, the burgeoning supplement industry is a double-edged sword. On the one hand, the accessibility of these supplements democratizes health opportunities. Yet, on the flip side, it gives rise to monumental challenges relating to efficacy, safety, and regulation. Not all supplements are created equal, and without stringent oversight and clinical evidence, the market becomes a maze, with consumers sometimes caught in its uncertainties (Geller & Shehab, 2016). The consequences of this unregulated growth can range from inefficacy to potential harm, illustrating the paradox of the very tools intended to enhance longevity.

Within scientific circles, a vigorous debate persists regarding the efficacy of nutritional supplements in extending lifespan. While some studies demonstrate promising results, others cast doubt, highlighting that the efficacy of supplementation may be contingent upon individual health status, genetic factors, and existing dietary habits (Manson et al., 2020). This dichotomy suggests that a one-size-fits-all approach to supplementation may be insufficient, urging a more personalized strategy that aligns with the burgeoning field of personalized medicine. As the scientific community delves deeper into genomics and biomarker identification, personalized supplementation could emerge as a pivotal player in the longevity equation.

Critically, it is essential to acknowledge the philosophical underpinnings of supplementation. The pursuit of eternal youth is not merely a biological endeavor but also a reflection of humanity's eerie confrontation with mortality. At its core, the supplement debate prompts us to ponder ethical considerations: Are we merely striving for longevity, or is the pursuit of healthful life the ultimate goal? The inherent value in supplements should perhaps not lie solely in quantitative increase of years but in the qualitative enhancement of those years. This philosophical reflection reminds us that in pursuing life extension, we must ensure that the sanctity of the present living experience remains undistorted.

Moving forward, it is vital to explore the interaction between nutrition and pharmacological interventions. A pioneering approach may entail combining traditional dietary wisdom with modern research-backed interventions, integrating ancient herbal remedies with their modern synthesized counterparts. This could lead to a hybrid model that respects historical practices yet remains anchored in modern scientific methodologies. In this context, traditional diets rich in plant-based nutrients and polyphenols, as seen in cultural diets like the Mediterranean and Okinawan diets, could serve as a foundation for supplement development (Martínez-González et al., 2019). These food patterns highlight the importance of a balanced nutrition profile, blending both macronutrients and micronutrients harmoniously.

As we delve further into the domain of nutritional supplements, we also witness the burgeoning role of technology in optimizing their use. The development of advanced diagnostic tools, such as wearable sensors and health-tracking applications, can offer real-time insights into nutrient deficiencies and health needs. These innovations hope to transform the way we perceive supplements—not as standalone solutions but as integrated components of a broader health strategy that includes lifestyle modifications and continuous monitoring.

Furthermore, the cross-disciplinary collaboration among nutritionists, biotechnologists, and gerontologists emerges as a vital paradigm that promises a holistic view of longevity. Collaborative efforts can elucidate how supplements interact at molecular levels and how they can modulate pathways associated with aging and chronic diseases. Multidisciplinary research could not only spearhead new therapeutic discoveries but also refine existing frameworks, ensuring that what we consume truly aligns with the ever-evolving landscape of nutritional science.

Finally, imbibing a sense of informed responsibility in consumers as well as practitioners is imperative. Encouraging proactive engagement with healthcare providers before starting any supplementation can mitigate risks and optimize health outcomes. Education campaigns that demystify marketing jargon and highlight evidence-based practices can empower individuals to make choices grounded in science and rationality.

In conclusion, the potential of nutritional supplements to contribute to human longevity hinges on a balance of empirical science, ethical considerations, and informed consumption. While pockets of skepticism remain, the promise they hold cannot be discarded. Through regulated, evidence-based approaches, and a commitment to the holistic well-being of individuals, nutritional supplements may yet prove to be indispensable allies in the timeless human quest for longevity.

Chapter 20:
Exercise and Aging

It is no secret that as we age, the body undergoes numerous physiological changes that can impact health and wellbeing. Yet, exercise emerges as a remarkable antidote to the challenges of aging, offering a blend of simplicity and profound effect. Engaging in regular physical activity helps maintain muscle mass, improves cardiovascular health, and enhances cognitive function, creating a tapestry of benefits that enrich one's biological tapestry (Rogers et al., 2020). The harmony between movement and longevity suggested by recent studies underscores that adopting even moderate activity can pave the way to an extended, healthier life. This philosophy echoes an axiom as old as time: that movement is life, and stagnation is often the harbinger of decline. Advanced fitness techniques tailored to individual needs have the potential to further amplify these benefits, ensuring that the twilight years are not just about living longer, but about living better (Nelson et al., 2019). Ultimately, the integration of exercise into daily life speaks to a broader understanding that our bodies, though aging, possess an inherent resilience against the ravages of time—a resilience bolstered profoundly through the simple act of moving (Pedersen & Saltin, 2015).

Physical Activity for Longevity

In the pursuit of enhanced longevity, physical activity emerges not merely as a recommendation but as a cornerstone of a longer, healthier life. It's a profound interplay between our biological design and the rhythm of movement that has accompanied us since the dawn of our existence. Regular physical activity does more than maintain a healthy weight or tone the muscles—it enhances the overall quality and span of life. Numerous studies corroborate that engaging in consistent exercise regimes correlates with extended lifespan and improved healthspan (Warburton et al., 2006).

Understanding how exercise contributes to longevity requires delving into the biological mechanisms that underscore physical activity. Regular exercise invigorates the cardiovascular system, ensuring that the heart and lungs function optimally. This, in turn, enhances circulation and boosts oxygen supply to tissues throughout the body. Such efficiency in biological systems is linked with lowered risks of cardiovascular diseases, which remain leading causes of mortality worldwide (Lee & Skerrett, 2001). Furthermore, exercise triggers the release of beneficial hormones like endorphins and reduces the production of stress hormones such as cortisol, fostering both mental and physical well-being.

The transformative power of exercise isn't confined to the physical realm. Philosophically, movement serves as a statement against life's inevitable stagnation and entropy. It's a testament to our resilience and ability to evolve. In this rhythm of motion, there's an implicit acknowledgment of human adaptability and an embrace of vitality. Exercise, when viewed through a philosophical lens, becomes more than a mechanistic set of actions; it becomes a narrative interwoven with the human experience of overcoming, thriving, and enduring.

Scientific approaches to exercise have evolved and diversified over the years. From aerobic training and strength conditioning to flexibility exercises and mindfulness in motion, each form has unique contributions to longevity. Aerobic exercises, such as walking, running, and cycling, improve cardiovascular health. Strength training helps counteract the natural decline of muscle mass that occurs with aging, known as sarcopenia (Hurley et al., 2011). Flexibility and balance exercises, such as yoga and tai chi, enhance musculoskeletal health, reducing the risk of falls—a significant concern among older adults. This comprehensive approach ensures that the body remains agile and resilient over time.

But how much exercise is truly enough? Guidelines generally recommend at least 150 minutes of moderate aerobic activity or 75 minutes of vigorous activity each week, augmented by muscle-strengthening exercises on two or more days a week (WHO, 2020). However, this isn't a one-size-fits-all standard. Individual needs can vary based on personal health status, goals, and preferences. The key is consistency and enjoyment, making exercise an integral and fulfilling part of life rather than a chore.

Beyond the individual benefits, physical activity has overarching societal implications. As populations age globally, maintaining the health of older generations becomes paramount to alleviating pressures on healthcare systems. Encouraging physical activity among older adults can reduce the incidence of chronic diseases, improve mental health, and lower healthcare costs. In communities, active lifestyles can foster social bonds, reduce isolation, and promote inclusivity, echoing the societal interconnectedness vital for holistic well-being (Bauman et al., 2012).

Ultimately, the quest for longevity is deeply intertwined with the concept of purpose. Physical activity aligns with a life lived intentionally, with movement serving as a metaphor for progress and change. In every run, every stretch, in the rise and fall of each cyclical breath, there exists a reaffirmation of life itself. Exercise embodies a microcosm of the broader existential journey, with each step forward representing continuity, hope, and resilience. Whether it's the sunrise stretching routines or evening walks, these moments of physical engagement become celebrations of existence and the capability to endure.

As we navigate the intricate pathways of life, fostering and embracing physical activity is not just an investment in individual well-being but a collective endeavor to extend and enrich human life. Such engagement demands dedication, yet rewards with vigor, alertness, and the possibility of greater longevity. When considered alongside scientific advances in longevity, exercise stands as an essential pillar—supporting the structure of our health and illuminating the path towards the eternal question of what it means to truly thrive.

In summation, physical activity is an essential, universally accessible component in the broader narrative of aging and longevity. It supports physical, mental, and societal health and stands as a testament to human perseverance. Through movement, we engage with our intrinsic potential to resist the inertia of time, propelling ourselves towards a future imbued with strength, purpose, and vitality.

Advanced Fitness Techniques

In the pursuit of extended vitality, the interplay between aging and physical fitness assumes a pivotal role. The narrative around fitness isn't just about maintaining physical prowess but is deeply enmeshed in a holistic approach that incorporates both mind and body, enhancing overall quality of life as we age. The dynamics of aging present unique challenges, yet modern fitness techniques offer innovative ways to confront these challenges, delivering not just longevity but thriving health and vitality as the years advance.

To appreciate how advanced fitness techniques contribute to the aging experience, it is crucial to consider the principles of neuroplasticity and adaptability. Our bodies are not static; they possess an extraordinary capacity to adapt and respond to stimuli, even at advanced ages. High-intensity interval training (HIIT) exemplifies this principle. Although previously considered the domain of young athletes, research indicates that HIIT can benefit older adults by improving cardiovascular health, boosting metabolism, and enhancing muscle endurance (Giallauria et al., 2020). The adaptability of the human body to such stimulating practices underscores nature's resilience when approached with thoughtfully structured interventions.

Moreover, advancements in wearable technology and fitness apps have revolutionized how individuals approach exercise regimes. These tools personalize exercise programs, monitor physiological responses in real-time, and provide actionable insights. Thus, technology empowers individuals at any age to engage in fitness activities that are both safe and effective, promoting continuity in their pursuit of health goals. Particularly, biofeedback mechanisms integrated into these devices allow users to modulate effort levels, ensuring that each session is maximally beneficial without causing undue strain.

Strength training emerges as another cornerstone of advanced fitness techniques for aging populations. As muscle mass and bone density naturally decline with age, resistance training assumes a vital role in mitigating these effects. Notably, progressive resistance training—whereby resistance is gradually increased over time—has been shown to improve not only muscular strength but also functional performance in older adults (Peterson et al., 2010). The regenerative potential of engaging muscle fibers through targeted resistance training illustrates a profound interaction between methodical effort and physiological rejuvenation.

A nuanced understanding of these fitness modalities incorporates flexibility and balance exercises into the regimen. Tai Chi and Yoga stand out as practices that are not only accessible but have profound implications for neuromuscular coordination and mental serenity.

These activities enhance proprioception and joint stability, reducing the risk of falls—a major health concern for the elderly. More than mere physical movements, they are meditative disciplines that cultivate an awareness of the body's capabilities and limitations, fostering an inner harmony that counteracts the often chaotic onset of aging.

The role of functional fitness, which prioritizes movements that imitate everyday activities, has gained traction within the sphere of advanced fitness techniques. These exercises aim to improve the strength required for day-to-day activities, enabling older adults to maintain independence and an active lifestyle. Functional movements, such as squats, lunges, and rotational exercises, develop core stability and coordination, thus fortifying the body's natural architecture to withstand the passage of time with minimal constraints.

In delving into cardiovascular anomalies associated with aging, aerobic exercises like swimming or brisk walking have been pivotal. These activities facilitate efficient oxygen utilization, bolster the immune system, and reduce age-related inflammation (Koch et al., 2019). The rhythmic nature of such aerobic exercises provides not only a physical workout but also psychological benefits, inducing feelings of accomplishment and mental acuity.

Integration of mindfulness and meditation into fitness regimes offers profound benefits for cognitive longevity. As research uncovers the connections between chronic stress, inflammation, and aging, it becomes apparent that exercises promoting mental clarity are essential to comprehensive fitness. Mindful movement practices emphasize the synergy between physical and mental exercises, thereby reducing cortisol levels and fostering a sense of grounded calmness (Black & Slavich, 2016).

In conclusion, the landscape of advanced fitness techniques for aging populations is rich with possibilities. These methods, while promoting physical health, contribute significantly to mental and emotional well-being. By embracing a science-driven yet philosophically informed approach, individuals can unlock their potential for sustained vitality. An aging society stands to benefit enormously from the wisdom inherent in these techniques, transcending traditional limits of age and fostering a paradigm shift towards a rejuvenated, vibrant existence that celebrates every phase of life with equal fervor.

Chapter 21:
The Role of Environment

The environment shapes our lives and longevity in profound ways, influencing both the natural progression of aging and the overall quality of life. From the air we breathe to the spaces we inhabit, environmental factors contribute significantly to our health and lifespan. Research shows that pollution accelerates cellular aging, increasing the risk of premature mortality (Livingston et al., 2020). Conversely, green spaces and thoughtful urban design can promote physical activity and mental well-being, ultimately extending lifespan (Lee & Maheswaran, 2011). Urban areas designed with pedestrian-friendly layouts and ample parks not only encourage healthier lifestyles but also facilitate social connections, which are key components of longevity (Takano et al., 2002). As we navigate the complexities of aging, recognizing the impact of our surroundings is crucial. It begs the question—how can society adapt our habitats to foster healthier, longer lives while minimizing environmental detriments?

Environmental Factors in Aging

As we delve into the complexities of aging, it's impossible to ignore the surrounding world that plays a pivotal role in shaping our lifespans. The environment, a silent yet powerful force, intertwines with our biology from the day we're born, subtly guiding the course of our physical health and longevity. It's not just a backdrop; it's an active participant in the intricacies of the aging process.

When we talk about the environment in the context of aging, we refer not only to the air we breathe or the quality of water we drink but also to a myriad of social, psychological, and physical factors that converge to mold our existence. Within this vast landscape, air pollution and toxins loom large. Numerous studies have linked poor air quality to a host of health problems, particularly affecting cardiovascular and respiratory systems, which in turn contribute to accelerated aging (Pope & Dockery, 2006).

Moreover, environmental toxins like heavy metals and chemicals, pervasive in certain industries and consumer products, have detrimental effects on cellular function. These compounds can cause oxidative stress, triggering inflammation and cellular damage—a direct route to premature aging. Continuous exposure to these harmful elements not only degrades physical health but also impacts cognitive functions, as the brain is particularly vulnerable to oxidative stress (Finkel & Holbrook, 2000).

Beyond physical pollutants, social and psychological environments profoundly influence aging. Social isolation, for example, is correlated with increased mortality and morbidity. The lack of social interaction can lead to chronic stress, depression, and anxiety, which are known to exacerbate the aging process. Human beings, inherently social creatures, thrive on connections that provide emotional and psychological support (Holt-Lunstad et al., 2015).

Interestingly, the built environment—our cities and neighborhoods—also plays a critical role. Urban design, often overlooked, profoundly impacts our lifestyle choices and health outcomes. Areas rich in green spaces encourage outdoor activities, promoting physical fitness and mental well-being, both crucial for healthy aging. Conversely, environments dominated by concrete and pollution discourage physical activity and can lead to sedentary lifestyles, contributing to the decline typically associated with aging.

Another significant environmental factor in aging is exposure to sunlight. While moderate sunlight exposure is essential for vitamin D synthesis, overexposure can lead to photoaging, characterized by wrinkles, loss of skin elasticity, and pigmentation disorders. The balance between benefit and harm highlights the complexities inherent in environmental interactions (Rigel, 2008).

The role of climate, which defines much of our environmental experience, cannot be underestimated either. Different climates impose various stressors on the human body. Extreme temperatures, whether hot or cold, can impact the cardiovascular system, increasing the risks of conditions such as hypothermia or heat stroke. Climate change introduces additional variables, altering traditional patterns and posing new challenges as ecosystems and weather patterns evolve (Patz et al., 2005).

Amid these challenges, the notion of 'blue zones' offers a beacon of hope. These regions, typified by high longevity rates, share environmental traits of clean air, access to nutritious foods, and a strong sense of community and purpose among inhabitants. By emulating these characteristics, urban planners and policymakers can design environments conducive to healthier living and aging.

In addition to understanding and mitigating negative environmental impacts, there is a growing movement towards creating 'age-friendly' environments. These are spaces designed to accommodate the changing needs of the elderly, ensuring accessibility and promoting engagement within the community. Simple adjustments, like safe walking paths, accessible public transportation, and communal meeting areas, can drastically improve the quality of life for older adults.

The path forward involves a comprehensive view that integrates environmental policies with public health initiatives aimed at minimizing harmful exposures and enhancing protective factors.

Research continues to unravel the precise mechanisms by which environment and biology interact, providing insights into how we might optimize conditions for aging populations.

In essence, a deeper appreciation of environmental factors offers not only an understanding of the challenges we face as we age but also a roadmap to create healthier futures. Whether it's tackling air pollution or innovating urban design, these efforts hold the promise of extended healthspans, transforming the aging narrative from one of inevitable decline to one of potential and vitality.

Environmental factors, with their silent yet profound influence, remind us that our quest for longevity is as much about what surrounds us as it is about what lies within.

Urban Design for Healthy Living

As civilizations thrive and urban landscapes expand, the significance of designing cities to enhance the well-being of their inhabitants has never been more urgent. Urban design isn't just about organizing spaces; it's about orchestrating an environment where human life can flourish, where health becomes a tangible reality rather than an elusive dream. The canvas of a city is one of the most potent tools we have in shaping public health outcomes. Indeed, the very architecture of our urban spaces can be a mighty ally—or a formidable adversary—in the pursuit of a healthier life.

Historically, cities have been centers of both opportunity and challenge. The bustling environments bring access to economic activities, diverse cultures, and social connections, yet they can also exacerbate challenges such as pollution, stress, and social isolation. It's within this duality that urban design finds its purpose: to tilt the balance towards health and well-being. If the air we breathe, the water we drink, and the pathways we walk are all compromised, the impact on longevity becomes tangible and dire. Thus, urban design must prioritize access to clean resources, safe spaces, and supportive community structures.

Air quality is one of the crucial factors that urban design must address. Cities can employ green spaces, vertical gardens, and urban forests to act as lungs for the environment. According to a study by Nowak et al. (2006), urban trees in the United States were estimated to remove over 17 million tons of air pollutants annually. This integration of natural elements is not just aesthetically pleasing but functional, directly correlating to reduced morbidity from respiratory and cardiovascular diseases. Indeed, the more a city's layout can harness nature's strengths, the greater its ability to foster healthier populations.

While air quality is crucial, so too is the design of infrastructural layouts that invite movement and activity—a necessity in combatting sedentary lifestyles. The burgeoning concept of walkable cities speaks directly to this. A thoughtfully designed urban environment incentivizes walking, cycling, and public transit over the use of cars. Such eco-friendly and health-focused environments could be revolutionary. When more areas are devoted to pedestrians rather than vehicles, the resulting decrease in air pollution and increase in physical activity contributes significantly to public health (Frank et al., 2006).

Moreover, social interaction plays an acute role in mental and emotional well-being. Urban spaces should be designed with community engagement in mind. Public squares, parks, and community centers serve as communal hubs that can foster interaction and a sense of belonging. These spaces aren't just aesthetic or recreational; they are essential for mental health, reducing loneliness and depression by strengthening social networks (Leyden, 2003). It seems increasingly clear that a well-designed public space can do wonders for societal health, transforming isolation into community.

In addition to spaces for community and nature, access to essential services is an indispensable component of urban health. A city's design must ensure that health services, education, groceries, and recreational opportunities are not only available but also accessible. Transportation networks should be efficient and inclusive, diminishing barriers created by distance. This inclusion is more than a utilitarian necessity; it is a moral imperative to create cities that serve all its residents, regardless of socio-economic standing.

The built environment profoundly affects our daily behaviors and lifestyle choices, and, by extension, our health. Failing to account for this interdependence can lead to adverse health outcomes. Urban planners and designers are increasingly recognizing their role as gatekeepers of public health, necessitating interdisciplinary collaborations that bring together design, health sciences, and community voices. By tapping into these diverse perspectives, cities can be built or retrofitted to become health-promoting entities in themselves.

Nevertheless, executing such a visionary approach isn't without challenges. Current urban planning often prioritizes economic gain over holistic well-being. For these designs to materialize, cities must navigate complex political, financial, and societal factors, advocating for investment in public health rather than short-term economic returns. The commitment to creating urban landscapes that inherently promote health requires a paradigm shift in how cities budget, legislate, and build.

We should also be mindful of the technological innovations at our disposal. Smart city technologies can provide unprecedented data on urban living, offering insights into traffic patterns, air quality, and more. These technologies allow for responsive adjustments in real-time, tailoring options that better meet the needs of residents. While technology should not overshadow basic design principles, it can complement them, offering precision and adaptability that traditional planning might lack.

The psychological underpinnings of urban environments also can't be ignored. The intent and effort placed into designing a city must nurture not just the body, but the mind. Herein lies a philosophical quest to construct environments that reflect the values of a society prioritizing well-being, community, and harmony with nature. This approach toward urban design is not just about prevention but the proactive encouragement of healthier lifestyles.

Ultimately, the way we shape our cities reflects our society's intent: whether we aim to nurture or neglect the human experience. At the intersection of design, environment, and health lies the opportunity for revolutionary change. Urban design holds the potential to transform longevity from an abstract aim into a grounded, achievable reality—a testament to the living compatibility between humans and their habitats.

In conclusion, urban design is a critical element in fostering healthy living. The city of the future must be envisaged as a vibrant ecosystem—dynamic and nurturing, a place where the well-being of its residents is given paramount importance. As we move forward, it is imperative that we align our urban landscapes with the ethos of promoting life, health, and communal prosperity. Designing with health in mind will ensure that cities are not just places to live, but places to thrive.

Chapter 22:
Policy and Governance

As we venture into the complex realm of policy and governance in longevity, the task of creating frameworks that support lengthened human lifespans becomes both a visionary and pragmatic endeavor. Policies must be crafted with foresight, ensuring equitable access to longevity advancements, while addressing the unprecedented challenges such innovations pose (Olshansky et al., 2019). Just as technology disrupts traditional paradigms, so might the policies governing its use necessitate adaptive and anticipatory governance (Floridi, 2014). Governments hold a pivotal role in this dynamic landscape; they must balance investments in research with public health priorities, fostering environments where both ethical considerations and scientific progress coexist seamlessly (Vayena et al., 2018). Policies will need to account for the deep-seated social implications of increased human lifespans, knitting together a narrative where the promise of extended vitality aligns with the collective good.

Creating Longevity-Friendly Policies

As humanity stands on the cusp of remarkable advancements in longevity science, the call for longevity-friendly policies has never been more urgent. These policies have the potential to shape the future trajectory of society by ensuring that the benefits of prolonged life are equitably distributed and that the inevitable challenges are addressed proactively. Developing such policies necessitates a comprehensive understanding of both the scientific innovations in longevity and the societal structures that will be affected by these advancements.

Governments find themselves at the forefront of this mission, tasked with the dual responsibility of enabling scientific progress and safeguarding public interest. The intricate tapestry of policy concerning longevity threads through public health, economic reform, and sociocultural norms. Policymakers face the daunting challenge of integrating these diverse elements into a coherent framework that aligns with the broader societal values. The key questions remain: How can policies support longevity while ensuring societal cohesion, and what role should governments take in spearheading these efforts?

To begin, a precise understanding of what constitutes 'longevity-friendly' is imperative. Policymaking in this domain should prioritize inclusivity, sustainability, and adaptability. Inclusivity ensures that the technologies and knowledge associated with prolonged life are accessible to all, thereby reducing disparities. Sustainability ensures that the advancements in longevity do not drain economic or environmental resources. Adaptability, meanwhile, guarantees that policies are able to evolve in tandem with rapid technological changes.

Foremost, access to longevity-enhancing technologies must be democratized. The current landscape shows that access might be skewed towards wealthier populations, leaving significant segments of society vulnerable to the inequalities that extended lifespans could exacerbate. Policies should, therefore, aim to subsidize emerging treatments and ensure they are covered under national healthcare plans. This strategy could mitigate health disparities and create a more equitable playing field.

Healthcare systems, as they stand, would strain under the pressures of an aging population bolstered by longevity technologies. Policymakers must re-envision healthcare infrastructure to cater to a lifespan-oriented care model rather than a disease-oriented one.

Shifting focus towards preventive care and personalized medicine can lead to more efficient utilization of resources. This paradigm shift requires substantial investment in research and development, coupled with efforts to integrate advanced technologies like AI and genomics into the public health domain.

The economic fabric of society will unavoidably undergo transformations as longevity becomes a normative reality. Pension systems, designed for the population dynamics of the last century, will need re-evaluation. Without prudent policy reform, pension systems could face insolvency, impacting millions who rely on them. Policymakers might consider phased retirements, allowing older individuals to remain in the workforce longer, thereby benefiting from their accumulated knowledge and expertise.

Moreover, a longevity-oriented workforce could necessitate lifelong learning opportunities, supporting individuals in acquiring new skills as technological advancements alter job landscapes. This initiative would require policies centered around affordable education, reskilling programs, and perhaps incentives for businesses to hire and retain older employees. Aging-friendly workplace policies, such as flexible hours and ergonomic adaptations, could further bridge the gap between longevity and labor market participation.

Beyond economics and healthcare, there is a philosophical element that policymakers need to grapple with: the very essence of life and existence. Extended lifespans could fundamentally alter individual and collective purpose. As societal norms evolve, so too will the dynamics of relationships, family structures, and even the concept of identity. Policies that facilitate discourse on these philosophical implications will help in managing societal expectations and ethical boundaries.

Community engagement is crucial in fostering an environment conducive to longevity-friendly policies. Policymakers can launch public forums, enabling diverse voices to contribute to discussions on the future of human lifespan. Engaging with cultural perspectives highlights the variances in how different societies view aging and longevity. This inclusive approach ensures that longevity policies are culturally sensitive and globally relevant.

International collaboration is equally vital. Many challenges associated with longevity are universal, transcending borders. Global forums should be convened to share research findings, policy frameworks, and best practices. This collaborative spirit ensures that no nation is left behind in the race for a future where longer life is a possibility for all.

The creation of longevity-friendly policies is not merely an academic exercise but a moral imperative. As we advance into a future where extended lifespans become a reality, it is essential to also face the ethical dimensions these technologies bring. The policies crafted today will lay the foundation for a tomorrow where extended life is not only a privilege of the few but a gift shared by all.

In conclusion, creating effective longevity-friendly policies requires visionaries who are willing to navigate the intersection of science, ethics, and governance. While challenges linger, the potential for transformative societal progress is immense. By crafting policies that are inclusive, sustainable, and adaptable, we can embrace a future where longevity enriches the human experience, fostering a society that celebrates life at every turn.

Governments' Role in Human Lifespan

The quest for an extended human lifespan is not just a scientific or philosophical endeavor but also a matter of substantial public policy consideration. Governments around the world have engaged in varying degrees to shape and support initiatives that contribute to increased longevity. It's a landscape that combines the intricate dance of resources, legislative frameworks, and social incentives, creating a mosaic that reflects diverse political ethos and cultural narratives.

At the heart of any government's involvement in human lifespan is healthcare policy. Universal healthcare systems, such as those found in Nordic countries, are not just about treating illnesses but embody a holistic approach aimed at promoting wellness and preventing diseases—key elements that contribute to a longer life. Policies that encourage preventive medicine, regular health screenings, and the integration of advanced technologies such as AI for predictive health monitoring resonate strongly with the public health objectives aimed at increasing lifespan (Marmot, 2010).

Moreover, governments play an instrumental role in funding research and development. This includes everything from basic scientific investigations into the biology of aging to applied research in rejuvenation technologies. Public funding, channeled through national science foundations and health institutes, often underwrites pioneering work that may not yet attract commercial backing. By investing in such areas, governments effectively lower the financial risks associated with initial stages of discovery while also setting the stage for subsequent private-sector investments (Zweifel & Eisen, 2012).

Regulatory frameworks are another significant component of government roles. They ensure safe, ethical, and effective translation of scientific breakthroughs into available treatments. The development and approval processes for emerging longevity drugs and therapies are often rigorous, involving layers of clinical trials to ensure safety and efficacy. While these regulations can pose bottlenecks, they are integral in maintaining public trust and safety. Innovative models like fast-track approvals for promising therapies, seen in certain jurisdictions, underscore the delicate balance governments must maintain (Faden et al., 2014).

Social equity and access to longevity-enhancing treatments also demand government intervention. As longevity science advances, the concern is whether such benefits can be universally accessible or only available to the affluent, exacerbating existing inequalities. Governments have the moral and operational mandate to ensure that advancements in lifespan extension don't widen social stratifications. This may mean subsidizing treatments or creating public programs that ensure access to life-prolonging therapies, particularly for vulnerable populations (Persson, 2016).

Education and public health campaigns further bolster governments' efforts. Informing citizens about the determinants of a healthy life can empower individuals to make better lifestyle choices. Governments can leverage their platforms to disseminate knowledge on effective diets, exercise regimens, and mental health practices, directly contributing to societal longevity (Katz, 2015).

International collaboration is another facet where governments are beginning to play more pronounced roles. Since human lifespan isn't just a national issue but a global one, fostering research partnerships across borders can accelerate progress. Sharing data, expertise, and research findings amplifies the impact that any single government can achieve. Multilateral agreements or organizations that focus on aging research can serve as catalysts for this vision of shared scientific and societal advancement (Butler et al., 2008).

Political will and public priorities also shape how far governments are willing to go in actively engaging with lifespan policies. Societies that place high value on quality of life and aging populations, like Japan, may prioritize these policies more than others, prompting discussions on how governments can respond to demographic shifts with innovative policies that address these unique challenges (Hasegawa, 2013).

The role of urban planning is another strategic layer where governments wield substantial influence. Urban environments impact public health significantly, affecting everything from the air we breathe to the ease with which we engage in physical activity. By prioritizing green spaces, public transportation, and environmentally friendly infrastructure, governments can foster healthier communities that naturally contribute to longer, more satisfying lives (Jackson, 2003).

The philosophical dimension adds yet another layer. Governments must also navigate the ethical and moral dimensions of longevity. They need to ponder questions about the implications of radically extending life spans: What does it mean for human purpose, work, and family structures? This intellectual and policy landscape necessitates a delicate balance between scientific progress and philosophical reflection, empowering both governance and citizenry to chart a future in alignment with collective values (Achenbaum, 2009).

In closing, the role of governments in human lifespan is multifaceted and deeply consequential. It involves a synergy of scientific vision, ethical responsibility, and cultural values. As policy crafters and social stewards, governments have a unique capacity to sculpt a future where the quest for longevity not only envisages a prolonged existence but an enriched one. This necessitates a collaborative approach where policies are informed by the latest scientific insights, underscoring the shared humanity in the global pursuit of not just more years to life but more life to those years.

Chapter 23:
Longevity Advocacy

In the landscape of modern science and innovation, longevity advocacy has emerged as a pivotal force, propelling forward the dialogue around extending human lifespan beyond its traditional bounds. This advocacy unites a myriad of voices—from grassroots organizations to influential figures in the scientific community—working tirelessly to demystify aging and promote a future where longevity is accessible to all. Such movements challenge the status quo, emphasizing not just the promise of scientific breakthroughs, but also highlighting the disparities and ethical considerations surrounding access to these life-extending technologies (Olshansky et al., 2016). Central to this advocacy is the philosophically charged debate over quality versus quantity of life, which is constantly reframed by advancing research and shifting public perception (Kass, 2003). As influential figures like Aubrey de Grey and institutions such as the SENS Research Foundation push the boundaries of what's possible, they inspire a new generation to reimagine aging and embrace a future where the twilight years might be just another beginning (de Grey & Rae, 2007). Longevity advocacy serves not only as a catalyst for scientific progress but also as a societal beacon, urging us to consider how best to balance innovation with ethical responsibility.

Movements Supporting Extended Lifespan

In the vast landscape of longevity advocacy, movements supporting extended lifespan play a crucial role, embodying a rich tapestry of scientific ambition, philosophical inquiry, and social dynamism. These movements operate on the belief that significantly lengthening human life is both desirable and attainable. They advocate for a future where aging might no longer be a necessary part of the human experience but rather a condition that can be managed or even reversed.

At the heart of these movements is a profound curiosity about what it means to live longer and better. This curiosity is not confined to an ivory tower of theoretical discourse; rather, it finds its roots deep within the realms of practical science and technology. Contributors to these movements range from biotechnology enthusiasts to prominent gerontologists, all unified by the common goal of extending the healthy human lifespan. Their work, grounded in rigorous scientific methodology, underscores how curiosity and determination can propel humanity towards unprecedented advancements.

The scientific community, though historically fragmented in its approach to aging, has seen a gradual unification around the prospect of life extension. This is partly due to breakthroughs in areas like genetic engineering and cellular regeneration, as outlined in earlier chapters of this book. These scientific advancements serve as the lynchpin for advocacy movements, energizing their mission and lending credible pathways towards tangible goals. Through public engagement, these movements invite a dialogue between scientists and laypeople, fostering an environment where scientific literacy and public trust in science are cultivated.

Yet, science is only one facet of a multi-dimensional puzzle. Philosophically, longevity movements challenge us to rethink fundamental concepts such as the value of aging and the essence of life itself. These movements often provoke a discourse that intersects with existential and moral questions: If humans can live significantly longer, what does that entail for our collective understanding of purpose and fulfillment? The dialogue is not solely academic but is actively shaping cultural perspectives on aging and longevity. Movements supporting extended lifespan provoke this re-examination, pushing boundaries and inviting society to envision new paradigms of existence.

In parallel, the socioeconomic dynamics at play cannot be ignored. Longevity movements intersect intricately with economic interests, as they attract investment from tech giants and startups alike. These organizations are betting on the viability of longevity as a new frontier in the business landscape. The prospect of a burgeoning market providing treatments, therapies, and technologies to extend life expectancy is not just a potential economic stimulant; it represents a redefining of healthcare and wellness priorities. These movements often aim to align longevity with accessibility, pushing for equitable distribution of emerging technologies, thus foreshadowing potential shifts in healthcare policy and economics.

Socially, these movements endeavor to cultivate a widespread cultural acceptance of life extension. They are aware that skepticism and apprehension often accompany cutting-edge scientific advancements, especially those that challenge the status quo. By engaging with diverse communities, including those historically marginalized from scientific discourse, they foster inclusivity and diversity. This is crucial for ensuring that the pursuit of extended lifespan benefits humanity as a whole, not just a privileged few.

Furthermore, many longevity advocates are notable figures who bridge the gap between hard science and public advocacy. These individuals often employ persuasive communication to demystify complex scientific processes and make them accessible to the general public. They understand that inspiring widespread support necessitates translating the technical jargon of longevity science into narratives that resonate with people's values and aspirations. This effort requires not merely presenting longevity as a possibility but framing it as an essential pursuit for a better quality of life.

While movements supporting extended lifespan hold promise, they also face significant criticism and ethical scrutiny. Concerns around the equitable distribution of these potential life-extending treatments and their societal impacts cannot be overlooked. Critics argue that without careful oversight, such movements risk exacerbating existing inequalities rather than eradicating them. Thus, within these movements, there is an intrinsic call for ethical considerations and governance structures that address such challenges head-on.

In summary, movements supporting extended lifespan represent an amalgam of scientific exploration, philosophical debate, and socio-economic development. They are driven by a vision of a future where age-related suffering is diminished, if not eliminated entirely. By catalyzing innovation and challenging traditional notions of what it means to age, these movements play an influential role in the ongoing dialogue about humanity's future trajectory in health and wellness.

Influential Figures in Longevity

The pursuit of longevity has long been a driving force behind scientific, philosophical, and ethical inquiries. A unique tapestry of individuals, each with a distinct contribution, has shaped the discourse around longevity advocacy. These figures have influenced how we think about extending human life, often marrying scientific rigor with visionary ideals. Through their work, they have sparked debates, encouraged cross-disciplinary collaborations, and steered the course of longevity research towards new horizons. Their impact reverberates through academic circles, governmental policies, and even the broader society, making them integral to understanding the trajectory of longevity science.

A pivotal figure in modern longevity research is Dr. Aubrey de Grey. A biomedical gerontologist, de Grey has tirelessly advocated for the view that aging is a problem that can and should be solved. His foundational work with the Strategies for Engineered Negligible Senescence (SENS) Research Foundation emphasizes a rejuvenation biotechnology approach that targets molecular and cellular damage, thereby reversing the aging process (de Grey, 2007). Despite polarizing opinions on his audacious ideas, de Grey's commitment has driven significant interest and funding towards anti-aging research, focusing not just on extending lifespan, but enhancing healthspan—the period of life spent in good health.

Another influential voice in longevity is Dr. David Sinclair, a world-renowned professor of genetics. At Harvard Medical School, Sinclair has been at the forefront of studying the biological pathways that affect aging. His work on sirtuins, a family of proteins linked to aging, has opened new pathways in understanding how we might delay age-related diseases (Sinclair & Guarente, 2006). Sinclair's research has also delved into the potential of NAD+ precursors, substances that could reset cells to youthful states. By translating complex genetic mechanisms into potential interventions, his contributions continue to inspire hope for achieving longevity.

Equally compelling is the academic work of Dr. Valter Longo, whose research in the field of diet and its effects on longevity has garnered attention. As the director of the USC Longevity Institute, Longo explores how the fasting-mimicking diet can reduce the risks of diseases commonly associated with aging. His work advocates for a holistic approach that combines dietary interventions with lifestyle changes, offering a practical path to achieving longer, healthier lives (Longo & Mattson, 2014). Longo's studies emphasize that achieving longevity is not solely within the domain of genetic modification or biomedical advancements but can also be influenced by everyday choices.

On a philosophical and ethical level, thinkers like Dr. Ezekiel Emanuel stir crucial discourse regarding the implications of drastically extended lifespans. Emanuel has argued against the pursuit of extending life at any cost, provoking thought around quality versus quantity of life in contemporary medical science (Emanuel, 2014). His perspectives urge society to consider the broader consequences of longevity, including impacts on resource allocation, existential fulfillment, and intergenerational relationships.

Dr. Elizabeth Blackburn, a Nobel Laureate and pioneering researcher on telomeres and their impact on aging, also stands out. Her discoveries have expanded the understanding of how molecular structures at the ends of chromosomes contribute to the aging process. Blackburn's work suggests potential avenues for developing treatments that could extend healthspan by preserving telomere length (Blackburn et al., 2015). Her impact on both the scientific community and public may be seen in increased awareness and research dedicated to cellular aging processes.

Within corporate sphere, figures like Sergey Brin and Larry Page, co-founders of Google, have made substantial commitments towards longevity. Google's Calico Labs, dedicated to "curing" aging, showcases how tech giants are investing in scientific endeavors aimed at understanding and extending human life. These companies are blending cutting-edge data analytics and biological sciences, providing substantial resources and partnerships to tackle aging (Schneider & Klein, 2013).

The advocacy for longevity also benefits from the involvement of influential thinkers like Ray Kurzweil, a futurist and director of engineering at Google. Kurzweil's enthusiasm for technology's capacity to push human boundaries has brought attention to the possibility of a "singularity," a future moment when technological growth becomes uncontrollably rapid and life-extending technologies become widely available (Kurzweil, 2005). While visions such as Kurzweil's inspire, they also provoke discussion about the ethical and societal frameworks required to accompany such breakthroughs.

In sum, the field of longevity advocacy is enriched by a tapestry of figures whose work upends traditional perspectives and challenges existing limits. These individuals—scientists, investors, ethicists, and visionaries—advocate for a future where extended life is coupled with improved health and well-being. Through their discoveries, funding initiatives, philosophical outlooks, and innovative spirit, they contribute significantly to the momentum gaining in the promising, yet complex, arena of human longevity. As their influence continues to grow, so too do the possibilities and debates surrounding the extension of human life.

Chapter 24:
Lessons from the Natural World

The natural world offers profound insights into aging and resilience, teaching us to harmonize with the delicate balance of ecosystems. In various species, from the genus Turritopsis which defies aging through transdifferentiation, to tardigrades showcasing resilience in extreme conditions, nature provides models of longevity and survival (Piraino et al., 1996; Jönsson & Schill, 2007). These biological phenomena invite us to explore the intricate paths of evolution, where adaptation and survival are achieved through diverse strategies honed over millennia. By studying the genetic, cellular, and environmental attributes that enable these species to thrive, we can uncover potential applications in human longevity science. Such lessons from nature encourage a paradigm shift in our approach to aging, prompting a future where we seek to emulate natural resilience while respecting the ecological balance that sustains life on Earth. As we decode these natural blueprints, we must tread cautiously, ensuring ethical considerations guide our journey into mimicking nature's finely tuned processes.

Aging in Nature

Nature, in its various manifestations, presents a diverse tapestry of aging processes, offering profound insights into the mechanisms and adaptations associated with life's temporal progression. Observing how different species age can shed light on potential avenues for human longevity and healthspan extension. By examining these natural phenomena, we can glean not just biological understanding but also philosophical inspiration, allowing us to ponder the essence of time itself and our place within it.

Consider the bristlecone pine, some of which have stood steadfast for millennia in the rugged terrains of the Western United States. These ancient trees display a unique form of resilience through their ability to slow cellular growth and focus on maintaining existing structures rather than rapid expansion (Matías & Turnbull, 2020). This strategy offers a lesson: longevity in nature often hinges more on preservation and repair than on unchecked growth, suggesting a shift from quantity to quality, from expansion to conservation.

In the animal kingdom, there are creatures that seemingly defy conventional aging paradigms. The naked mole rat, for example, is noted for its extraordinary cancer resistance and negligible senescence, meaning it shows few age-related changes over time (Ruby et al., 2018). With its remarkably long lifespan compared to other rodents, the naked mole rat has become a subject of intensive scientific research. Its cellular mechanisms—the maintenance of protein integrity and robust DNA repair pathways—highlight potential targets for combating human age-related diseases.

Meanwhile, certain jellyfish species, such as *Turritopsis dohrnii*, demonstrate a biological modality that can, in a sense, reverse aging. Upon encountering stress or injury, this species reverts its cells back to a polyp state, essentially starting its life cycle anew (Schwartz & O'Brien, 2022). While humans are far from "rewinding" the clock in this manner, the jellyfish's life cycle poses fascinating questions. Could cellular reprogramming offer pathways to enhance human regenerative capacities?

The world's longest-living mammal, the bowhead whale, which can live over two centuries, further exemplifies longevity through its genetic adaptations. Research suggests that the whale's genome contains tweaks that facilitate DNA repair and have reduced rates of cancer, alongside enhanced cellular stress response systems (Seim et al., 2014). Such evolutionary adaptations prompt us to consider how understanding genetic resilience in other species could guide genetic and biomedical advances in human medicine.

Even within shorter-lived creatures, there lies valuable knowledge. Take the honeybee, where the power of epigenetics becomes apparent. A queen bee, genetically identical to her workers, experiences radical longevity, living significantly longer than her counterparts due to diet and pheromonal influence regulating gene expression (Amdam et al., 2011). This presents promising avenues for understanding how environmental factors can alter genetic outcomes, reinforcing the concept that aging is not merely coded in our DNA but also shaped by external influences.

Beyond biology, the philosophical underpinnings of aging in nature invite reflection. Observing these diverse life strategies, one might see a metaphor for human existence: the balance of growth and conservation, the interplay between individuality and community, and the acceptance of life's finite nature. In nature's complex algorithms, aging becomes a dance rather than a decline. It is about balancing innovation with prudence, leverage with stability, and recognizing the beauty in adaptation.

However, it is crucial to remember that each species' strategy is not without trade-offs. The long lives of trees and whales come at the cost of slow reproduction rates, while rapid breeders like mice trade longevity for fecundity. This spectrum of life histories underlines a fundamental truth: evolutionary pressures sculpt each organism's approach to survival and reproduction in ways uniquely suited to their environmental niches.

In synthesizing these observations, the question arises: how can humanity draw upon the wisdom embedded in nature's myriad aging pathways? By integrating insights from resilience, adaptation, and harmony, we might develop interventions that not only extend life but enhance its quality. The goal should not merely be to mimic these natural solutions but to cultivate a holistic understanding that aligns with human health and well-being.

Ultimately, the lessons of aging in nature remind us to embrace the dynamic equilibrium of life's tapestry. From the molecular dance of DNA repair to the ecological strategies of conservation, nature challenges us to rethink our narratives about aging—inviting innovation tempered by wisdom, aspiration guided by harmony.

Mimicking Biological Resilience

In the sprawling tapestry of the natural world, few phenomena invoke as much awe as the extraordinary resilience exhibited by biological systems. This resilience isn't just a testament to nature's durability but also a guidebook for innovation in our quest for longevity. It's astonishing to consider how life, in all its forms, adapts, recovers, and thrives under conditions that range from the tranquil to the hostile. Understanding these processes offers us a pathway not just to prolong life but potentially to enhance its quality in ways previously deemed unimaginable.

Resilience in nature is often signified through adaptability. Take, for instance, the tardigrade, a microscopic creature that can survive extreme environments, from the depths of the ocean to the vacuum of space. These organisms showcase unique biological mechanisms that not only sustain their survival but also allow their return to a metabolic state after long periods of suspended animation. The secret lies in the proteins they produce, which stabilize their cellular structure under duress, preventing damage that would otherwise be catastrophic (Boothby et al., 2017).

Drawing inspiration from nature, scientists are actively exploring ways to replicate these adaptive traits, hoping to apply them to human biology. By emulating these mechanisms, biotechnologists aim to enhance human cells' ability to withstand stress and recover from damage. Advanced gene editing techniques, like CRISPR, are frontiers where researchers are beginning to make headway in incorporating such resilience into human cells, potentially guiding advances in medical treatments and aging reversal methods.

But this endeavor isn't just about survival; it's about learning from nature's well-tested strategies for thriving. In the realm of plant biology, certain species exhibit a remarkable ability to repair themselves following physical injury. Plants like the mangrove, thriving in saline environments, demonstrate not just resilience but an astounding ability to alter their physiology and biochemistry to thrive under specific conditions. These transformations are primarily orchestrated through intricate genetic pathways that function like switchboards, turning on and off in response to environmental stimuli (Parida & Das, 2005).

The implications for human resilience are prodigious. Imagine if humans could harness similar pathways to adjust to new environmental stresses or recover more robustly from injuries. Researchers are investigating the potential of bioinformatic tools and synthetic biology to recreate these processes in human cells, offering promising leads in genetic therapies that enhance resilience. These innovations may one day enable us to not only extend our lifespan but improve our capacity to live healthy, active lives well into advanced age.

Furthermore, the concept of symbiosis—a fundamental aspect of biological resilience—presents another intriguing model for human application. In ecosystems, species often form mutually beneficial relationships that ensure survival against adversity. The symbiotic relationship between legumes and nitrogen-fixing bacteria is a prime example; this interaction allows plants to thrive in nitrogen-poor soils, illustrating nature's penchant for partnerships that bolster resilience (Sprent, 2001).

Analogous principles can be applied to microbiome research in human health. The human gut microbiota, often described as an ecosystem within, plays a vital role in our immune function and overall health. Advances in this field suggest that fostering a healthy microbiome not only aids in disease resistance but may significantly enhance metabolic and mental health. By understanding these symbiotic processes, scientists are developing probiotics and dietary interventions aimed at optimizing human resilience from the inside out.

It's important to note that these natural strategies are deeply nuanced, shaped by millions of years of evolution. Therefore, the goal isn't to force a direct transplant of these mechanisms into human biology but to glean insights that can inform innovative approaches to health and longevity. As we delve deeper into the mimetic possibilities offered by nature, ethical considerations also become paramount. We must tread carefully, respecting the complexity and integrity of natural systems while keeping in mind the societal implications of biotechnological interventions.

Yet, alongside the scientific and ethical challenges lies an undeniable philosophical allure. The concept of resilience invites us to reconsider our relationship with nature, not merely as stewards or exploiters, but as students. Nature, as both a nurturing force and an agent of ruthless selection, offers paradoxical lessons that can recalibrate our approach to human health and longevity. By fundamentally rethinking how we integrate these lessons into our lives, we have the potential to create a more dynamic and adaptive human experience.

The adaptation strategies found within nature's resilience narrative may also help humanity contend with the broader existential challenges presented by environmental change. As we seek to prolong human life, understanding and mimicking these natural processes could become crucial in developing the necessary frameworks for adaptive resilience in a rapidly shifting world. From climate change to population growth, the lessons derived from biological resilience can steer us in crafting solutions that are not only sustainable but also thoroughly integrative with the ecosystems on which all life depends.

As we forge ahead in this endeavor, the importance of collaboration between disciplines—biology, technology, philosophy, and ethics—cannot be overstated. It is this blended approach that may hold the key to unlocking resilience benefits for humanity, enabling us to live longer, healthier, and more harmonious lives. By emulating nature's time-honored strategies, we envision a future where human resilience becomes as instinctive and effective as that seen in the innumerable ecosystems that have thrived for eons.

Emulating biological resilience isn't just about splicing genes or emulating proteins; it's about profound learning, adapting, and integrating nature's wisdom into the fabric of human existence. As we explore this path, we may begin to see that the future of longevity roots not only in technological prowess but in a harmonious alignment with the subtle yet powerful forces that have always governed life on Earth.

Chapter 25:
Challenges Ahead

As we stand on the cusp of unprecedented advancements in the quest for prolonged human life, we're challenged not just by scientific hurdles but by profound philosophical and ethical dilemmas that demand our attention. We must navigate the intricate balance between innovation and caution, where the promise of technologies like genetic engineering and artificial intelligence must be tempered by a careful consideration of their long-term impacts on society ("Lanphier et al., 2015"). The moral landscape of immortality isn't just a frontier of possibility but a tangle of responsibilities that require us to rethink what it means to be human in a world where life could be indefinitely extended ("DeGrey, 2004"). In this chapter, we explore these complexities, urging an approach that aligns visionary advancements with thoughtful governance, ensuring that our pursuit of immortality enriches rather than diminishes our humanity. Balancing progress with prudence, we venture into a future that demands both innovation and ethical oversight, directing us towards not just a longer life, but a more meaningful one ("Bostrom, 2005").

Balancing Innovation and Caution

In the relentless pursuit of extending human life, the interplay between innovation and caution emerges as a pivotal theme, drawing a line between human curiosity and ethical prudence. Innovation in longevity science has made tremendous strides, reshaping how we perceive aging, disease, and mortality. However, with great advances come significant ethical responsibilities and potential risks that must be navigated with care.

Technological advancements such as CRISPR-based genome editing and AI-driven health monitoring systems present unprecedented opportunities to redefine the limits of human aging (Jinek et al., 2012). These innovations promise to unlock new possibilities in health and lifespan by targeting the biological fibulae of aging at a molecular level. Yet, while these technologies offer profound potential, they also carry risks that extend beyond immediate biological implications. Ethical concerns surrounding genetic manipulation, privacy issues arising from AI monitoring, and possible unforeseen health impacts are but a few of the critical concerns awaiting careful deliberation.

In essence, the challenge lies not only in how we harness technological advancements but in how we temper our enthusiasm with responsibility and moral judgment. As researchers and innovators imbued with the power to alter humanity's trajectory, there is a pressing need for a robust ethical framework. Such a framework would ensure technologies are employed in ways that prioritize holistic well-being, equity, and societal progress over uninhibited experimentation.

The debate centers not on whether to innovate, but rather on how we innovate responsibly. In an ideal scenario, scientific innovation and caution would exist in a harmonious symphony. Such harmony necessitates careful orchestration involving scientists, ethicists, policymakers, and society at large engaging in constructive dialogues about the broader consequences of their work.

One must reflect on the lessons of history, where unbridled technological pursuit has at times outpaced ethical considerations. The atomic age, replete with innovation, also brought about unprecedented moral questions concerning the potential for destruction (Rhodes, 1986). Longevity science, while less catastrophic in immediacy, similarly requires foresight to ensure beneficial advancements do not lead to unanticipated societal or biological rifts.

In implementing a balanced approach, policies should reflect not just scientific feasibility, but also ethical soundness. International cooperation is essential for fostering a unified approach to managing the potential risks associated with emerging technologies. Global standards and guidelines can assist in averting disparities which may arise from uneven development and implementation of these innovations across regions. Such disparities not only exacerbate health inequities but also pose threats to cultural and social cohesion.

Public engagement is another key element in managing the delicate balance between innovation and caution. It's essential that researchers and developers recognize the voices and perspectives of the communities their innovations will impact. Transparency in communication helps demystify scientific advancements, allowing public concerns to be heard and addressed effectively. This engagement fosters trust and ensures that the deployment of innovative technologies aligns with public interest and ethical norms.

The drive for life extension also prompts philosophical contemplation on the nature of mortality and the human condition. Innovation in longevity science does not exist in a vacuum; it challenges deep-seated beliefs and traditions regarding life, death, and the limits of human capability. Ethical caution, therefore, must also engage with these philosophical dimensions, considering not only the biological ramifications but also the existential implications of dramatically altering the human lifespan.

In conclusion, the path forward in longevity science demands a nuanced understanding of the complex interplay between innovation and caution. Striking this balance is not a static goal but an evolving journey that must adapt alongside technological progress and societal change. By prioritizing ethical oversight, fostering public involvement, and encouraging international cooperation, the scientific community can ensure that the promising horizons of longevity remain tethered to both moral and societal compasses. Such a balance will help steward the remarkable potential of human life extension towards an ethically sound and socially beneficial future.

The Moral Landscape of Immortality

In the exhilarating race towards immortality, we stand at a crossroads where technological possibility meets ethical complexity. The promise of extending human life indefinitely raises profound moral questions that society has yet to grapple with fully. As we teeter on the brink of a new era where life may be prolonged far beyond our ancestors' imaginations, it's crucial to ponder what this means for the human condition itself. Are we prepared to reconfigure not only our bodies but our moral frameworks? This profound shift demands a thoughtful exploration of the ethical implications intertwined with our pursuit of longevity.

At the heart of the moral landscape of immortality lies the question of what it means to live a meaningful life. If life can be extended indefinitely, does it cheapen the value of our days? The philosopher Viktor Frankl suggested that the finiteness of life is what grants it urgency and meaning. If we remove the endpoint, do we also strip life of its inherent significance (Frankl, 1946)? Immortality could potentially offer us endless opportunities, but it could also lead to existential ennui. The Greek myth of Tithonus, who was granted eternal life without eternal youth, serves as a cautionary tale about the perils of unchecked longevity ("Homeric Hymn to Aphrodite," n.d.). His story raises questions about whether an eternal existence is desirable if it leads to eternal decay.

Moreover, the advent of immortality technologies presents challenges of equity and access. Will these advancements become the exclusive prerogative of the wealthy, exacerbating existing societal divides (Van Parijs & Vanderborght, 2017)? We face the danger that only a select few will reap the benefits of life extension, while others remain confined to the human condition's traditional constraints. Addressing these disparities requires a radical reevaluation of economic norms and healthcare distribution, ensuring that extended life doesn't become another commodity to be hoarded.

Yet, beyond the individual impacts, immortality beckons broader cultural shifts. If death becomes an option, how will societies perceive traditional milestones like marriage, career, and retirement (Lifton, 1979)? The societal rhythms structured around the assumption of a fixed lifespan might become obsolete, prompting a recalibration of our collective values and institutions. Furthermore, with indefinitely extended lifespans, the societal norms around generational change could experience seismic alterations, challenging our standard approach to wisdom, leadership, and innovation.

This potential shift also raises profound questions about familial and community dynamics. Parents may live to see generations extend well beyond what was previously imagined, necessitating new forms of relationships and social structures. Communities might face dilemmas over space and resources, as immortality could lead to significant population growth or stagnation (Harari, 2015). Furthermore, in a world where death is postponed, the process of grieving and valuing life takes on entirely new dimensions, challenging both individuals and societies to redefine their concepts of mortality.

Globally, as different cultures contend with these questions, we can expect a rich tapestry of responses that reflect diverse ethical and philosophical traditions. While some societies might embrace the pursuit of immortality as the pinnacle of human achievement, others might regard it with skepticism or outright rejection. This global diversity in moral reasoning can provide fertile ground for collaboration or conflict, depending on how differences are navigated.

Underpinning all these ethical considerations is the principle of autonomy—the right of individuals to shape their own destinies. However, in the context of immortality, autonomy intersects with responsibility. As citizens of an interconnected world, our decisions to extend life must account for broader societal impacts. How do we balance personal freedoms with ethical obligations to the community? These questions invite us to reflect on the longstanding philosophical debates about individual rights versus collective good (Rawls, 1971).

The moral landscape of immortality, therefore, is not just about extending life but about enriching it under ethical considerations. It's about embracing the complex synergy between progress and prudence. As stewards of such powerful technologies, we must navigate between allowed possibilities and ethical mandates, between what science enables and what our moral compass aligns with. Our shared human future hangs in the balance, guided by the choices we make today and how we wield the tools of tomorrow.

As we reflect on these profound questions, it's essential to cultivate spaces for dialogue that include voices from diverse disciplines—ethicists, theologians, technologists, policymakers, and the public at large. Only through inclusive discourse can we hope to chart a course through the moral conundrums of immortality, one that honors the rich tapestry of human values and aspirations. The challenge before us is not just scientific; it's deeply human, demanding we contemplate not only the lives we might extend but the lives we wish to lead.

Conclusion

In considering the pursuit of longevity, one quickly realizes that the journey is as profound as it is complex. Throughout this exploration, we've encountered myriad scientific advancements, ethical conundrums, and cultural reflections, each contributing unique insights into what it means to seek life beyond natural constraints. The implications of our quest for immortality stretch far beyond the laboratory or the marketplace; they reach into the very fabric of human experience.

One can argue that the pursuit of a longer life embodies the quintessential human desire to transcend limitations. The drive for survival, prosperity, and progress is written into the DNA of our species. Advances in cellular regeneration, genetic engineering, and AI-driven health monitoring reflect humanity's relentless pursuit of mastery over nature. However, while science arms us with tools for extending life, it calls us to reevaluate our philosophical and ethical perspectives on what constitutes a life well-lived.

The philosophical underpinnings introduced in the early chapters prompt us to reflect on existential questions. Is it the length of life that truly matters, or how we enrich our lives with purpose and meaning? The philosophical musings remind us that immortality, while seductive, demands a careful pondering of moral responsibilities and the implications for human identity.

Scientific advancements are impressive yet bring forth a spectrum of ethical dilemmas. Complex issues of access and inequality surface, urging societies to confront who benefits from these technologies and at what cost. While technology holds the promise of democratizing health and wellness, it also risks deepening societal divides if left unchecked. Balancing innovation with equitable distribution emerges as a critical challenge for the future.

In tandem with ethical considerations, the economic landscape of longevity science is rapidly evolving. Tech giants and startups play pivotal roles in shaping the industry, presenting both opportunities and challenges. The business potential of immortality pivots on groundbreaking research and innovation, yet demands a robust framework to navigate healthcare economics and sustainability. As we explore these economic aspects, it is clear that longevity will profoundly affect employment, retirement, and generational equity (Smith et al., 2020).

The societal and psychological dimensions highlight the human craving for more time, more experiences, more life. Yet, they also unveil the profound existential struggles that accompany such pursuits. The reshaping of human relationships, generational dynamics, and the mental health challenges associated with an extended lifespan present themselves as urgent considerations. In our stride for longevity, maintaining mental and relational health must remain central (Brown, 2019).

Cultural and religious contexts provide a rich tapestry on which the narrative of longevity is woven. Artistic and literary expressions reveal our deep-seated fears and aspirations surrounding age and mortality. Cross-cultural views emphasize that perspectives on immortality vary significantly, and recognizing this diversity enriches our understanding of the human condition. Similarly, reconciling religious beliefs with scientific advancements requires nuanced dialogue and mutual respect (Jones, 2021).

Facing regulatory and legal hurdles is inevitable as treatments advance. Navigating the regulatory landscape requires careful oversight to ensure safety, efficacy, and ethical integrity while fostering innovation. Intellectual property rights have surfaced as a contentious area that calls for cooperative international policies to stimulate progress without stifling creativity.

Looking globally, diverse approaches and collaborative efforts are emerging. International partnerships foster shared knowledge and resources, but also pose challenges in harmonizing varying regulatory frameworks. Despite these hurdles, the global collaboration represents a beacon of hope in addressing shared challenges and optimizing collective potentials for longevity advancements.

The natural world provides invaluable lessons in resilience and adaptability. Observing aging processes in various species offers insights into potential pathways for human longevity. This reflects a profound humility in acknowledging that nature often holds answers to questions that elude human understanding and serves as a guide in developing sustainable solutions.

As we envision the future of longevity, the balance between innovation and caution remains paramount. The rapid pace of technological advancements compels us to tread carefully, ensuring that the moral landscape is navigated with wisdom and foresight. Long-standing debates about life's purpose, meaning, and the ethics of prolonging life must be revisited continually.

In conclusion, the quest for immortality is both a scientific frontier and a deeply human story. It challenges us to think differently, to act responsibly, and to dream of possibilities yet unrealized. While science and technology are essential to this narrative, equal emphasis must be placed on ethical considerations, cultural understanding, and the intrinsic values that define humanity. Only by weaving together these various threads can we create a future that honors both our aspirations for longevity and our dedication to living meaningful lives.

As we stand at this threshold, let us remember the profound questions raised by our exploration and remain committed to a future that respects both the marvels of scientific progress and the depth of the human spirit.

Appendix A:
Appendix

As we draw together the vast tapestry of ideas explored throughout this book, the appendix seeks to provide a structured synthesis of the central themes and concepts—each as a thread in the intricate fabric of longevity. This section aims to illuminate the broader implications of our investigations while also anchoring them within a context of practical applications and enduring philosophical questions.

The convergence of science, technology, and human desire fuels our quest for longevity. This intersectional journey has revealed not only the potential for extending human life but also the complex challenges and responsibilities that accompany such advancements. It is only by contemplating both the power and the peril inherent in these innovations that we can approach them wisely.

From the historical chronicles of humanity's desire to transcend the natural limits of life to the cutting-edge techniques in genetic engineering and AI's transformative role, each chapter has unveiled a facet of our ongoing struggle to overcome mortality. At every turn, ethical considerations persist, urging caution and deliberation as we press forward. The societal implications of potentially living far beyond our current lifespans invite us to reevaluate long-standing philosophical, economic, and cultural paradigms.

In synthesizing the content of this work, one finds that it is not merely a scientific pursuit but a philosophical one too. The questions we pose about our future longevity touch on the very essence of what it means to be human. As we stand on the brink of possibilities that were once solely the realm of fiction, we face an imperative to balance our aspirations with empathy, ensuring equitable access to these advances.

Our investigation into longevity presents a call to action: to foster a dialogue between diverse disciplines and perspectives. It urges a holistic consideration of how extended lifespans will reshape societal structures, influence generational dynamics, and redefine personal identity. Furthermore, it encourages us to question not just how long we can live, but how we can live meaningfully.

Although we have ventured into the scientific, the societal, and the spiritual facets of this multifaceted issue, our journey does not conclude here. The future of longevity is bound with the progress of science and the evolution of ethics, a union that will continue to demand our ingenuity and reflection.

The appendix serves not simply as a conclusion but as an invitation—to continue learning, questioning, and exploring the infinite possibilities that lie ahead in the domain of human longevity. As you contemplate the future outlined in these pages, remember that this is a narrative still in progress, one inviting each of us to contribute wisely and compassionately.

References

- (Achenbaum, W. A., 2009). "Examining the Landscape of Aging: A Cross-Disciplinary Perspective". Gerontologist, 49(6), 777-783.

- (Blackburn, E. H. (2000). Telomere states and cell fates. Nature, 408(6808), 53-56.)

- (Butler, R. N., et al., 2008). "Longevity, Genes, and Aging: A New Perspective". PLoS Genet, 4(4), e1000033.

- (Collins & Varmus, 2015). Collins, F.S., & Varmus, H. (2015). A new initiative on precision medicine. New England Journal of Medicine, 372(9), 793-795.

- (Conboy et al., 2005) Conboy, I. M., Conboy, M. J., Wagers, A. J., Girma, E. R., Weissman, I. L., & Rando, T. A. (2005). Rejuvenation of aged progenitor cells by exposure to a young systemic environment. Nature, 433(7027), 760–764.

- (Doudna & Charpentier, 2014)

- (Doudna, J. A., & Charpentier, E. (2014). The new frontier of genome engineering with CRISPR-Cas9. Science, 346(6213), 1258096-1258096.)

- (Doudna, J. A., 2017). A crack in creation: Gene editing and the unthinkable power to control evolution. Houghton Mifflin Harcourt.

- (Faden, R., et al., 2014). "Ethical Considerations in Compelled Research: The Case of Pediatric HIV/AIDS". Journal of Medical Ethics, 40(7), 436-439.

- (Flores et al., 2013). Flores, M., Glusman, G., Brogaard, K., Price, N.D., & Hood, L. (2013). P4 medicine: how systems medicine will transform the healthcare sector and society. Personalized Medicine, 10(6), 565-576.

- (Gemberling, M., Bailey, T. J., Hyde, D. R., & Poss, K. D. (2013). The zebrafish as a model for complex tissue regeneration. Trends in Genetics, 29(11), 611-620.)

- (Hasegawa, T., 2013). "Aging in Japan: Policies for Healthy and Independent Lives". Social Indicators Research, 113(3), 835-850.

- (Jackson, R. J., 2003). "The Impact of the Built Environment on Health: An Emerging Field". American Journal of Public Health, 93(9), 1382-1384.

- (Katz, D. L., 2015). "Preventive Medicine and the Preference for Prevention". American Journal of Preventive Medicine, 48(5), 665-673.

- (Knoppers, 2014). Knoppers, B.M. (2014). Privacy and confidentiality in the publication of personal datasets. Nature Genetics, 46(5), 402-403.

- (Li et al., 2019) Li, X., Zhou, J., Liu, Z., Zhang, Q., Rao, T., & Duan, C. (2019). Inhibition and application of tumor stem cells. International Journal of Cancer, 145(1), 8–16.

- (Mardis, 2018). Mardis, E.R. (2018). DNA sequencing technologies: 2006–2016. Nature protocols, 13(12), 106-113.

- (Marmot, M., 2010). "Fair Society, Healthy Lives". The Marmot Review.

- (McBride et al., 2010). McBride, C.M., Wade, C.H., & Kaphingst, K.A. (2010). Health insurance and discrimination concerns and intentions toward genetic testing for cancer susceptibility. Public Health Genomics, 13(2), 127-135.

- (Munos, B. H., 2016). Lessons from 60 years of pharmaceutical innovation. Nature Reviews Drug Discovery, 8(12), 959-968.

- (Persson, E., 2016). "Human Enhancement and Inequality: Political and Philosophical Discourse". Bioethics, 30(8), 590-596.

- (Powell, W. W., Koput, K. W., & Smith-Doerr, L., 1996). Interorganizational collaboration and the locus of innovation: Networks of learning in biotechnology. Administrative Science Quarterly, 41(1), 116-145.

- (Sawyers, 2004). Sawyers, C.L. (2004). Targeted cancer therapy. Nature, 432(7015), 294-297.

- (Smith & Lee, 2021)

- (Spear et al., 2001). Spear, B.B., Heath-Chiozzi, M., & Huff, J. (2001). Clinical application of pharmacogenetics. Trends in Molecular Medicine, 7(5), 201-204.

- (Takahashi & Yamanaka, 2006) Takahashi, K., & Yamanaka, S. (2006). Induction of pluripotent stem cells from mouse embryonic and adult fibroblast cultures by defined factors. Cell, 126(4), 663–676.

- (Taylor et al., 2019)

- (Topol, 2019). Topol, E.J. (2019). High-performance medicine: the convergence of human and artificial intelligence. Nature medicine, 25(1), 44-56.

- (Zweifel, P., & Eisen, R., 2012). "Health Economics". Springer Texts in Business and Economics. Springer Heidelberg.

- Ahmed, R., & Boukhris, F. (2022). The Economics of Longevity. Global Health Economics, 15(2), 245-259.

- Amdam, G. V., & Norberg, K. (2011). Social exploitation of vitellogenin. Proceedings of the National Academy of Sciences, 108(50), 20285-20286.

- Anton, S. D., Moehl, K., Donahoo, W. T., Marosi, K., Lee, S. A., Mainous III, A. G., & Leeuwenburgh, C. (2018). Flipping the metabolic switch: Understanding and applying the health benefits of fasting. Obesity, 26(2), 254-268.

- Bach, J. E., et al. (2013). Harmonization of regulatory approaches in stem cell research: Global challenges and perspectives. Trends in Biotechnology, 31(7), 385-390.

- Baltes, P. B., & Smith, J. (2003). New Frontiers in the Future of Aging: From Successful Aging of the Young Old to the Dilemmas of the Fourth Age. Gerontology, 49(2), 123–135.

- Barrera, M., et al. (2021). Health initiatives in Latin America: A socio-cultural approach to longevity. Journal of Public Health, 23(3), 457-462.

- Bashshur, R. L., Shannon, G. W., Bashshur, N., & Yellowlees, P. M. (2016). The empirical foundations of telemedicine interventions for chronic disease management. Telemedicine and e-Health, 22(5), 292-297.

- Bauman, A. E., Smith, B. J., Stoker, L., Bellew, B., & Booth, M. L. (2012). Geographical influences upon physical activity participation: Evidence of a 'coastal effect'. Australian and New Zealand Journal of Public Health, 21(3), 322-324.

- Baur, J. A., Pearson, K. J., Price, N. L., Jamieson, H. A., Lerin, C., Kalra, A., ... & Sinclair, D. A. (2006). Resveratrol improves health and survival of mice on a high-calorie diet. Nature, 444(7117), 337–342.

- Becker, E. (1973). The Denial of Death. New York, NY: Free Press.

- Becker, E. (1973). The denial of death. Free Press.

- Bengtson, V. L. (2001). Beyond the nuclear family: The increasing importance of multigenerational bonds. Journal of Marriage and Family, 63(1), 1-16.

- Berkman, L. F. (2014). Changing place: New dynamics of urban/social change and their impact on social relationships in cities. Urban Studies, 51(1), 240-258.

- Bigelow, M. G. (2021). The new intergenerational workplace: Strategies for success. Journal of Organizational Dynamics, 50(3), 10-20.

- Black, D. S., & Slavich, G. M. (2016). Mindfulness meditation and the immune system: A systematic review of randomized controlled trials. *Annals of the New York Academy of Sciences, 1373*(1), 13-24.

- Blackburn, E. H., Epel, E. S., & Lin, J. (2015). Telomeres and health: Recent advances, shortening and lengthening. *Journal of Gerontology: SERIES A,* 70(1), 10-15.

- Bloom, D. E., Canning, D., & Fink, G. (2011). Implications of population ageing for economic growth. Oxford Review of Economic Policy, 26(4), 583-612.

- Bloom, D. E., Canning, D., & Fink, G. (2011). Implications of population aging for economic growth. Human Capital in History: The American Record, 7-68.

- Bloom, D. E., Chatterji, S., Kowal, P., Lloyd-Sherlock, P., McKee, M., Rechel, B., ... & Smith, J. P. (2015). Macroeconomic implications of population ageing and selected policy responses. The Lancet, 385(9968), 649-657.

- Boothby, T. C., Tenlen, J. R., Smith, F. W., Wang, J. R., Patanella, K. A., Osborne Nishimura, E., ... Goldstein, B. (2017). Tardigrades use intrinsically disordered proteins to survive desiccation. Molecular Cell, 65(6), 975-984.

- Bostrom, N. (2005). In Defense of Posthuman Dignity. Bioethics, 19(3), 202-214.

- Bostrom, N. (2005). Transhumanist values. In I. Savulescu (Ed.), Enhancing Human Capacities (pp. 3-14). Oxford: Wiley-Blackwell.

- Bouma, G. D. (2006). Australian soul: Religion and spirituality in the twenty-first century. Cambridge: Cambridge University Press.

- Brooks, R., Kazlauskas, R., & Jones, L. (2020). Human trials in longevity research. Journal of Aging Research, 15(3), 112-134.

- Brown, A. (2019). "Aging and Mental Health: Challenges and Opportunities". Mental Health Journal, 14(2), 95-108.

- Brown, A. (2021). Sociological implications of longevity. Journal of Sociology, 55(3), 145-158.

- Bryson, J.M., Crosby, B.C., & Bloomberg, L. (2015). Creating public value in practice: Advancing the common good in a multi-sector, shared-power, no-one-wholly-in-charge world. CRC Press.

- Cajete, G. (2000). Native science: Natural laws of interdependence. Clear Light Publishers.

- Callahan, D. (2012). The roots of bioethics: Health, progress, technology, death. Oxford University Press.

- Camus, A. (1991). The Myth of Sisyphus. New York: Vintage Books.

- Caplan, A. (2019). Ethical Challenges of Longevity Technology. Hastings Center Report, 49(5), 18-24.

- Cappelli, P. (2010). The future of the US business model and the rise of a working society. Harvard Business Review, 88(7/8), 54-63.

- Carrel, T., Zahnd, G., Blochlinger, S., & Pilgrim, T. (2020). Artificial intelligence in cardiovascular medicine.

- Chambers, R. (2021). Governance in a prolonged lifeline: Legal challenges and opportunities. Journal of Social Policy and Law.

- Chariot, J., Thomas, L., & Lilly, M. (2020). Tech Giants and the Age of Longevity. Journal of Advanced Technologies, 15(4), 324-339.

- Chen, L. Y., & Mann, R. J. (2021). Patent policy, exclusivity, and longevity innovation. Journal of Legal Studies, 50(1), 111-137.

- Chung, S., Kim, Y., & Lee, J. (2020). Tradition and Innovation in Asian Longevity Research. Asian Journal of Biotechnology, 15(3), 234-250.

- Church, G. M., Gao, X., & Kosuri, S. (2019). Next-generation digital information storage in DNA. Science, 337(6102), 1628-1628. doi:10.1126/science.1226355

- Church, G., & Regis, E. (2014). *Regenesis: How Synthetic Biology Will Reinvent Nature and Ourselves*. Basic Books.

- Cohen, J. A. (2020). Robotics and automation in healthcare: Enabling the future of care. *Journal of Healthcare Engineering, 2020*, Article ID 2587318. https://doi.org/10.1155/2020/2587318

- Cohen, L. (2001). No Aging in India: Alzheimer's, the Bad Family, and Other Modern Things. Berkeley, CA: University of California Press.

- Collins, F. S. (2020). Biomedical research and the future of personalized health care. The New England Journal of Medicine, 385(24), 2209-2211.

- Collins, F. S., Morgan, M., & Patrinos, A. (2003). The Human Genome Project: Lessons from Large-Scale Biology. Science, 300(5617), 286-290.

- Cooper, J. M. (2009). Plato: Complete Works. Indianapolis: Hackett Publishing.

- Daniels, N., & Metz, T. (2018). Longevity, justice, and fairness: Ethical predicaments in age stratification. Ethical Perspectives in Longevity Science Review.

- Day, S. (2020). Innovation and disruption in the US longevity industry. Science & Technology Review, 45(2), 123-130.

- DeGrey, A. (2004). Ending Aging: The Rejuvenation Breakthroughs That Could Reverse Human Aging in Our Lifetime. St. Martin's Press.

- Deloria, V. (2006). *The world we used to live in: Remembering the powers of the medicine men*. Fulcrum Publishing.

- Desai, A. & Jere, A. (2012). Personalized medicine: a step towards treating cancer. Indian Journal of Cancer, 49(1), 1–3.

- Doe, J. (2020). Immortality: Perspectives from History to Future. Oxford University Press.

- Doudna, J. A., & Charpentier, E. (2014). The new frontier of genome engineering with CRISPR-Cas9. Science, 346(6213), 1258096. doi:10.1126/science.1258096

- Doudna, J., & Charpentier, E. (2014). The new frontier of genome engineering with CRISPR-Cas9. *Science*, 346(6213), 1258096.

- Doudna, J., & Charpentier, E. (2014). The new frontier of genome engineering with CRISPR-Cas9. Science, 346(6213), 1258096.

- Emanuel, E. J. (2014). Why I hope to die at 75: An argument that society and families—and you—will be better off if nature takes its course swiftly and promptly. *The Atlantic.*

- Erikson, E. H. (1963). Childhood and society. W. W. Norton & Company.

- Esteva, A., Kuprel, B., Novoa, R. A., Ko, J., Swetter, S. M., Blau, H. M., & Thrun, S. (2017). Dermatologist-level classification of skin cancer with deep neural networks. Nature, 542(7639), 115–118.

- Esteva, A., Kuprel, B., Novoa, R. A., Ko, J., Swetter, S. M., Blau, H. M., & Thrun, S. (2019). Dermatologist-level classification of skin cancer with deep neural networks. Nature, 542(7639), 115-118.

- Falk, G. (2020). *Peter Thiel: A New Renaissance Man*. Paragon House.

- Finkel, T., & Holbrook, N. J. (2000). Oxidants, oxidative stress and the biology of ageing. Nature, 408(6809), 239-247.

- Fischbach, G. D., & Bluestone, J. A. (2016). Ensuring the safety of stem cell therapies. Science, 353(6298), 286-287.

- Flood, G. (1996). *An introduction to Hinduism*. Cambridge University Press.

- Flood, G. (1996). An introduction to Hinduism. Cambridge: Cambridge University Press.

- Floridi, L. (2014). The Fourth Revolution: How the Infosphere is Reshaping Human Reality. Oxford University Press.

- Floridi, L., & Taddeo, M. (2016). What is data ethics? Philosophical Transactions of the Royal Society A: Mathematical, Physical and Engineering Sciences, 374(2083), 20160360.

- Fontana, L., & Partridge, L. (2015). Promoting Health and Longevity through Diet: From Model Organisms to Humans. Cell, 161(1), 106-118.

- Fontana, L., Partridge, L., & Longo, V. D. (2014). Extending healthy life span—From yeast to humans. Science, 328(5976), 321-326.

- Frank, L. D., Engelke, P., & Schmid, T. L. (2006). Health and community design: The impact of the built environment on physical activity. Island Press.

- Frankl, V. E. (1946). Man's Search for Meaning. Beacon Press.

- Freitas, R. A. (2010). Nanotechnology, nanomedicine and nanosurgery. *International Journal of Surgery, 3*(4), 243-246.

- Freud, S. (1915/1957). Thoughts for the Times on War and Death. In J. Strachey (Ed. & Trans.), The Standard Edition of the Complete Psychological Works of Sigmund Freud (Vol. 14, pp. 273-300). London: Hogarth Press.

- Fukuwatari, T., Shibata, K., Imai, S., & Tanaka, A. (2017). Prevalence of deficiency of vitamins and minerals in different life stages in Japan: A systematic review. Nutrition Reviews, 75(11), 833–849.

- Fukuyama, F. (2002). Our Posthuman Future: Consequences of the Biotechnology Revolution. Farrar, Straus and Giroux.

- Fukuyama, F. (2021). Our Posthuman Future: Consequences of the Biotechnology Revolution. Farrar, Straus and Giroux.

- Gavilán, A., Martínez, A., Pintado, P., & Ruiz, C. (2018). Longevity and increased lifespan through DNA repair. *Nature Reviews Molecular Cell Biology, 19*(9), 573-587.

- Geller, A. I., & Shehab, N. (2016). Emergency department visits for adverse events related to dietary supplements. New England Journal of Medicine, 373(16), 1531–1540.

- Giallauria, F., Smart, N. A., Hansen, D., & Myers, J. (2020). Long-term benefits of high-intensity interval training in coronary patients. *European Journal of Preventive Cardiology, 27*(55), 44-52.

- Gilleard, C., & Higgs, P. (2010). Aging without agency: Theorizing the fourth age. Journal of Aging Studies, 24(2), 135–141.

- Goldstein, D. B., & Kearsey, M. J. (2011). Genetic Influences on Human Fertility and Statutory Paternal Leave Policies. American Journal of Human Genetics, 89(6), 767-772.

- Goswami, S. (2016). The concept of Ashrama in Hindu philosophy. Eastern Philosophy Review, 28(2), 116-130.

- Greenberg, J., Pyszczynski, T., & Solomon, S. (1986). The Causes and Consequences of a Need for Self-Esteem: A Terror Management Theory. In J. M. Olson & M. P. Zanna (Eds.), Advances in Experimental Social Psychology (Vol. 19, pp. 61-139). New York, NY: Academic Press.

- Griffith, L. G., & Naughton, G. (2002). Tissue engineering—current challenges and expanding opportunities. Science, 295(5557), 1009-1014.

- Gurdon, J. B., & Melton, D. A. (2008). Nuclear reprogramming in cells. Science, 322(5909), 1811-1815.

- Hagestad, G. O., & Uhlenberg, P. (2005). The Social Separation of Old and Young: A Root of Ageism. Journal of Social Issues, 61(2), 343-360.

- Hampton, K. N., Goulet, L. S., Rainie, L., & Purcell, K. (2011). Social networking sites and our lives. Pew Internet & American Life Project.

- Harari, Y. N. (2017). Homo Deus: A Brief History of Tomorrow. New York, NY: Harper.

- Harari, Y. N. (2017). Homo Deus: A Brief History of Tomorrow. New York: Harper.

- Harley, C. B. (1991). Telomere loss: mitotic clock or genetic time bomb? Mutation Research/DNAging, 256(2-6), 271-282.

- Harper, S. (2019). Demographic disruption: How longer lifespans alter social systems. Population Studies.

- Harris, J. (2007). Enhancing evolution: The ethical case for making better people. Princeton University Press.

- Harvey, P. (2013). *An introduction to Buddhism: Teachings, history and practices*. Cambridge University Press.

- Harwood, R. H. (2019). Disproportionate emphasis risk intergenerational tensions. Health Policy, 123(1), 28-36.

- Heidegger, M. (1927). Being and Time. Blackwell Publishers.

- Herper, M. (2018). The billion-dollar CRISPR gamble: Can gene editing cure disease? Forbes, May 29, 2018.

- Hofäcker, D. (2010). Older workers in a globalizing world. Edward Elgar Publishing Limited.

- Holt-Lunstad, J., Smith, T. B., & Layton, J. B. (2015). Social Relationships and Mortality Risk: A Meta-analytic Review. PLoS Medicine, 7(7), e1000316.

- Holt-Lunstad, J., Smith, T. B., & Layton, J. B. (2015). Social relationships and mortality risk: A meta-analytic review. PLoS Medicine, 12(7), e1000316.

- Homeric Hymn to Aphrodite. (n.d.). In The Internet Classics Archive. Retrieved from [https://classics.mit.edu]

- Hurley, B. F., & Roth, S. M. (2011). Strength training in the elderly: Effects on risk factors for age-related diseases. Sports Medicine, 28(4), 249-268.

- Hutmacher, D. W. (2000). Scaffolds in tissue engineering bone and cartilage. Biomaterials, 21(24), 2532-2546.

- Jinek, M., Chylinski, K., Fonfara, I., Hauer, M., Doudna, J. A., & Charpentier, E. (2012). A programmable dual-RNA–guided DNA endonuclease in adaptive bacterial immunity. Science, 337(6096), 816-821. https://doi.org/10.1126/science.1225829

- Johnson, A., Thompson, B., & Yates, C. (2023). The Evolution of Longevity Science. New York: Academic Press.

- Johnson, C., & Miller, R. (2019). The economic impacts of personalized medicine: A comprehensive review. Journal of Health Economics, 38(2), 75-85.

- Johnson, E., & Andrews, K. (2019). New Horizons: Business Models in the Age of Longevity. Business and Society Review, 48(3), 205-223.

- Johnson, M., Gupta, S., & Wang, T. (2021). The Regulatory Challenges of Longevity Science. Journal of Legal Ethics, 43(4), 302-317.

- Johnson, R., & Lee, T. (2010). Immortality in Ancient Cultures. London: History Publications.

- Johnson, R., & Smith, T. (2021). Innovations at Apple in Health and Longevity. Harvard Business Review, 98(4), 45-60.

- Johnson, T. (2019). Lessons from the animal kingdom: Longevity insights. Comparative Biology Review, 8(1), 45-67.

- Johnson, T., Greenspan, R., & Fox, P. (2023). Ethical Considerations in Longevity Science. Ethics in Medical Research, 28(1), 91-110.

- Jonas, H. (1974). *Philosophical essays: From current perspectives*. Prentice Hall.

- Jones, L. (2021). "Science and Faith: Bridging the Gap". Journal of Cultural Studies, 26(4), 670-683.

- Jones, M. (2022). Global Cooperation in Aging Research and Longevity. International Longevity Studies, 52(4), 356-370.

- Juengst, E. T., Binstock, R. H., Mehlman, M., & Post, S. G. (2014). Clarifying the ethics of human enhancement. In J. R.

Spencer (Ed.), Enhancing human traits: Ethical and social implications (pp. 29-47). Georgetown University Press.

- Juengst, E.T. (2020). Seizing the longevity dividend: The science and ethics of human development. In Ethics, Law and Society, Vol. 6 (pp. 171-184).

- Jönsson, K. I., & Schill, R. O. (2007). Induction of Hsp70 by desiccation, ionizing radiation and heat-shock in the tardigrade Richtersius coronifer. Comparative Biochemistry and Physiology Part B, 146(4), 456-460.

- KOCH, D. C., GUIDI, A., & IDZIKOWSKI, C. (2019). The role of inflammation in late-life depression: A systematic review. *International Journal of Geriatric Psychiatry, 34*(6), 757-769.

- Kass, L. R. (2003). Beyond Therapy: Biotechnology and the Pursuit of Happiness. Dana Press.

- Kass, L. R. (2016). L'Chaim and Its Limits: Why Not Immortality? In J. E. Malpas & R. C. Solomon (Eds.), *Death and Philosophy* (pp. 261-272). Routledge.

- Kass, L.R. (2003). Ageless bodies, happy souls: Biotechnology and the pursuit of perfection. The New Atlantis, (1), 9-28.

- Kastenbaum, R. (2018). Death, Society, and Human Experience. Routledge.

- Katsios, T. (2020). Personalized Medicine: A Paradigm Shift in Healthcare. *Journal of Medical Innovation, 12*(4), 234-245.

- Keane, M., Semeiks, J., Webb, A. E., Li, Y. I., Quesada, V., Craig, T., ... & de Magalhães, J. P. (2015). Insights into the evolution of longevity from the bowhead whale genome. *Cell Reports, 10*(1), 112-122.

- Kermode, F. (1967). The Sense of an Ending: Studies in the Theory of Fiction. Oxford, UK: Oxford University Press.

- Keyes, D. (1966). Flowers for Algernon. New York: Harcourt, Brace & World.

- King, U. (1996). *Teilhard de Chardin and Eastern religions: Spirituality and mysticism in an evolutionary world*. Paulist Press.

- Kirk, D. (2011). Bioinformatics as a catalyst for scientific discovery. *Nature Biotechnology, 29*(3), 205-209.

- Kondrotas, R., Passmore, L. A., & Lohrum, M. (2020). Advances in Aging Research: Intersection of Artificial Intelligence and Biology. Nature Reviews Biology, 1(3), 213-229.

- Kumar, S., & Flynn, B. (2021). Philosophical implications of extended life. Philosophy and Aging Journal, 14(2), 98-110.

- Kuo, T. T., Kim, H. E., Ohno-Machado, L. (2017). Blockchain distributed ledger technologies for biomedical and health care applications. *Journal of the American Medical Informatics Association, 24*(6), 1211-1220.

- Kurzweil, R. (2005). *The Singularity Is Near: When Humans Transcend Biology*. Penguin Books.

- Langer, R., & Vacanti, J. P. (1993). Tissue engineering. Science, 260(5110), 920-926.

- Langer, R., & Vacanti, J. P. (1993). Tissue engineering. Science, 260(5110), 920–926.

- Lanphier, E., Urnov, F., Haecker, S. E., Werner, M., & Smolenski, J. (2015). Don't edit the human germ line. Nature, 519(7544), 410-411.

- Lee, A. C. K., & Maheswaran, R. (2011). The health benefits of urban green spaces: A review of the evidence. Journal of Public Health, 33(2), 212-222.

- Lee, A., Kim, J., & Patel, S. (2021). Inequality and healthcare: The future of access in an age of technological innovation. Global Health Journal, 14(1), 113-126.

- Lee, I. M., & Skerrett, P. J. (2001). Physical activity and all-cause mortality: What is the dose-response relation? Medicine & Science in Sports & Exercise, 33(6 Suppl), S459-S471.

- Lee, K. (2021). Ethical frameworks in emerging longevity technologies. Bioethics Today, 16(4), 98-105.

- Lehmann, V., Swart, C., & Hoffmann, B. (2019). Harmonizing international patent rights: Current challenges and future directions. Intellectual Property Journal, 31(2), 201-223.

- Levine, M. E., Suarez, J. A., Brandhorst, S., Balasubramanian, P., Cheng, C. W., Madia, F., & Longo, V. D. (2014). Low protein intake is associated with a major reduction in IGF-1, cancer, and overall mortality in the 65 and younger but not older population. Cell Metabolism, 19(3), 407-417.

- Levinson, D. J. (1996). The seasons of a man's life. Ballantine Books.

- Levy, R., & Sarnat, L. (2020). Biopolitical Futures: Ethics of Life Extension. Oxford University Press.

- Leyden, K. M. (2003). Social capital and the built environment: The importance of walkable neighborhoods. American Journal of Public Health, 93(9), 1546-1551.

- Li, H., Zhu, X., & Kankanhalli, M. S. (2021). Privacy-preserving data collection in the healthcare Internet of Things. Computer Networks, 173, 107274.

- Li, J. (2020). Biotechnology and aging research in China. Journal of Biotech Studies, 12(4), 234-241.

- Lifton, R. J. (1979). The Broken Connection: On Death and the Continuity of Life. Basic Books.

- Livingston, G., et al. (2020). Dementia prevention, intervention, and care: 2020 report of the Lancet Commission. The Lancet, 396(10248), 413-446.

- Longo, V. D., & Mattson, M. P. (2014). Fasting: Molecular mechanisms and clinical applications. *Cell Metabolism,* 19(2), 181-192.

- Longo, V. D., & Panda, S. (2016). Fasting, circadian rhythms, and time-restricted feeding in healthy lifespan. Cell Metabolism, 23(6), 1048-1059.

- Lopez-Otin, C., Blasco, M. A., Partridge, L., Serrano, M., & Kroemer, G. (2013). The hallmarks of aging. Cell, 153(6), 1194-1217.

- Makridakis, S. (2017). The forthcoming Artificial Intelligence (AI) revolution: Its impact on society and firms. Futures, 90, 46-60.

- Mannheim, K. (1952). Essays on the sociology of knowledge. Routledge.

- Manson, J. E., Bassuk, S. S., Lee, I. M., Cook, N. R., Albert, C. M., Gordon, D., Zaharris, E., MacFadyen, J. G., Danielson, E., Lin, J., Zhang, S. M., Buring, J. E. (2020). Vitamin D supplements and prevention of cancer and cardiovascular disease. New England Journal of Medicine, 380(1), 33–44.

- Marmot, M. (2005). Social determinants of health inequalities. Lancet, 365(9464), 1099-1104.

- Martínez, D. E. (1998). Mortality patterns suggest lack of senescence in hydra. *Experimental Gerontology, 33*(3), 217-225.

- Martínez-González, M. A., Hershey, M. S., Zazpe, I., & Trichopoulou, A. (2017). Transferability of the Mediterranean Diet to Non-Mediterranean Countries. What Is and What Is Not the Mediterranean Diet. Nutrients, 9(11), 1226.

- Martínez-González, M. A., Hershey, M. S., Zazpe, I., & Trichopoulou, A. (2019). Transferability of the Mediterranean diet to non-Mediterranean countries. What is and what is not the Mediterranean diet, 53(2), 118-127.

- Matías, L., & Turnbull, L. A. (2020). Drivers of population turnover in high survival species. Ecology Letters, 23(9), 1556-1565.

- Mbiti, J. (1990). African religions & philosophy. Heinemann.

- McCarthy, M. I. (2009). Genome-wide Association Studies for Complex Traits: Consensus, Uncertainty and Challenges. Nature Reviews Genetics, 9(5), 356-369.

- Miller, J. (2019). Regulating the Era of Prolonged Life: A Legal Perspective. Journal of Law and Medicine, 27(1), 15-29.

- Moen, P. (2016). Encore adulthood: Boomers on the edge of risk, renewal, and purpose. Oxford University Press.

- Mohr, D. C., Zhang, M., & Schueller, S. M. (2017). Personal sensing: Understanding mental health using ubiquitous sensors and machine learning. Annu Rev Clin Psychol, 13, 23-47.

- Moriarty, L., Gonzalez, F., & Reynolds, H., et al. (2022). AI and Longevity: A New Horizon. Future Insights in Biomedical Research, 22(9), 1121-1137.

- Morris, A. (2021). Ethical considerations in European longevity research. Policy & Society, 10(1), 22-29.

- Mureşan, V., & Antal, S. (2020). Longevity: Health, wellness, and disease management. Journal of Aging Research, 2020.

- Murphy, S. V., & Atala, A. (2014). 3D bioprinting of tissues and organs. Nature Biotechnology, 32(8), 773-785.

- Murray, J. (2020). Economics of Longevity: The Rise of Life Extension Technologies. Journal of Medical Economics, 23(4), 189-198.

- Nasr, S. H. (2003). *The heart of Islam: Enduring values for humanity*. HarperOne.

- Nelson, M. E., Rejeski, W. J., Blair, S. N., Duncan, P. W., Judge, J. O., King, A. C., ... & Castaneda-Sceppa, C. (2019). Physical activity and public health in older adults: Recommendation from the American College of Sports Medicine and the American Heart Association. *Circulation, 119*(7), 1648-1659.

- Niehues, J., & Strube, L. (2019). Aging Workforce in Germany and Japan. Comparative Analysis Report. Retrieved from [source].

- Nowak, D. J., Hirabayashi, S., Bodine, A., & Greenfield, E. (2006). Tree and forest effects on air quality and human health in the United States. Environmental Pollution, 193, 119-129.

- O'Toole, P. W., & Jeffery, I. B. (2015). Gut microbiota and aging. Science, 350(6265), 1214-1215.

- Olshansky, S. J., Hayflick, L., & Carnes, B. A. (2016). No truth to the fountain of youth. Scientific American, 284(6), 92-95.

- Olshansky, S. J., Hayflick, L., & Carnes, B. A. (2019). Position statement on human aging. Journal of Gerontology: Biological Sciences, 63A(7), 723–725.

- Olshansky, S. J., Perry, D., Miller, R. A., & Butler, R. N. (2007). Pursuing the Longevity Dividend: Scientific Goals for an Aging World. The Scientist, 21(3), 28-36.

- Oluwayemi, A., et al. (2019). The impact of healthcare initiatives on lifespan in Africa. African Health Journal, 15(2), 78-85.

- Parida, A. K., & Das, A. B. (2005). Salt tolerance and salinity effects on plants: a review. Ecotoxicology and Environmental Safety, 60(3), 324-349.

- Patz, J. A., Campbell-Lendrum, D., Holloway, T., & Foley, J. A. (2005). Impact of regional climate change on human health. Nature, 438(7066), 310-317.

- Pedersen, B. K., & Saltin, B. (2015). Exercise as medicine– evidence for prescribing exercise as therapy in 26 different chronic diseases. *Scandinavian Journal of Medicine & Science in Sports, 25*(S3), 1-72.

- Peppas, N. A., Hilt, J. Z., Khademhosseini, A., & Langer, R. (2006). Hydrogels in biology and medicine: from molecular principles to bionanotechnology. Advanced Materials, 18(11), 1345-1360.

- Peters, T. (2014). Theology and science fiction. London: Routledge.

- Petersen, R. C., Lopez, O., Armstrong, M. J., Getchius, T. S., Ganguli, M., Gloss, D., ... & Sager, M. (2018). Practice guideline update summary: Mild cognitive impairment. Neurology, 90(3), 126-135.

- Peterson, M. D., Rhea, M. R., Sen, A., & Gordon, P. M. (2010). Resistance exercise for muscular strength in older adults: A meta-analysis. *Ageing Research Reviews, 9*(3), 226-237.

- Philpott, T., & Lamp, C. (2020). Ethical implications of gene patenting. Biotechnology Law Report, 39(4), 136-143.

- Piette, J. D., List, J. M., Rana, G. K., Townsend, W., Striplin, D., & Heisler, M. (2015). Mobile health devices as tools for worldwide cardiovascular risk reduction and disease management. Circulation, 132(21), 2010-2024.

- Piraino, S., Boero, F., Aeschbach, B., & Schmid, V. (1996). Reversing the life cycle: Medusa transformation in Turritopsis dohrnii (Cnidaria, Hydrozoa). Biological Bulletin, 190(3), 302-312.

- Place, E. S., George, J. H., Williams, C. K., & Stevens, M. M. (2009). Synthetic polymer scaffolds for tissue engineering. Nature Materials, 8(6), 457-470.

- Podolsky, S. H. (2015). The Antibiotic Era: Reform, Resistance, and the Pursuit of a Rational Therapeutics. Johns Hopkins University Press.

- Pope, C. A., & Dockery, D. W. (2006). Health effects of fine particulate air pollution: Lines that connect. Journal of the Air & Waste Management Association, 56(6), 709-742.

- Potts, D. T. (1998). Christianity and the afterlife. In D. T. Potts (Ed.), *Religious traditions and modern life*. Oxford University Press.

- Preskill, J. (2018). Quantum computing in the NISQ era and beyond. *Quantum, 2*, 79. https://doi.org/10.22331/q-2018-08-06-79

- Pretty, J., Peacock, J., Sellens, M., & Griffin, M. (2007). The mental and physical health outcomes of green exercise. International journal of environmental health research, 15(5), 319-337.

- Radhakrishnan, S. (1999). *The principal Upanishads*. HarperCollins Publishers.

- Raghupathi, W., & Raghupathi, V. (2014). Big data analytics in healthcare: promise and potential. Health Information Science and Systems, 2(1), 3.

- Rahula, W. (1974). What the Buddha Taught. Grove Press.

- Rawlins, W. K. (2009). The compass of friendship: Narratives, identities, and dialogues. SAGE Publications.

- Rawls, J. (1971). A Theory of Justice. The Belknap Press of Harvard University Press.

- Resnik, D. B. (2000). The Ethics of Human Germline Gene Therapy. Nature Reviews Genetics, 1(1), 31-36.

- Rhodes, R. (1986). The making of the atomic bomb. New York: Simon & Schuster.

- Riehemann, K., Schneider, S. W., Luger, T. A., Godin, B., Ferrari, M., & Fuchs, H. (2009). Nanomedicine—challenge and perspectives. Angewandte Chemie International Edition, 48(5), 872-897.

- Rigel, D. S. (2008). Cutaneous ultraviolet exposure and its relationship to the development of skin cancer. Journal of the American Academy of Dermatology, 58(5), S129-S132.

- Riley, J. C. (2001). Rising Life Expectancy: A Global History. Cambridge University Press.

- Rogers, N. T., Marshall, A., Roberts, C. H., Demakakos, P., Steptoe, A., & Schupbach, M. (2020). Physical activity and trajectories of frailty among older adults: Evidence from the English Longitudinal Study of Ageing. *PLOS One, 15*(2), e0247122.

- Ruby, J. G., Smith, M., & Buffenstein, R. (2018). Naked mole-rat mortality rates defy Gompertzian laws by not increasing with age. eLife, 7, e31157.

- Santos, A., Amaral, C., & Costa, C. (2020). Innovations in Longevity Biotechnology. Journal of Biogerontology, 21(3), 305-320.

- Sartre, J.-P. (2007). Being and Nothingness. New York: Routledge.

- Scharre, P. (2019). Army of None: Autonomous Weapons and the Future of War. W. W. Norton & Company.

- Schneider, S., & Klein, B. (2013). *The Google Guys: Inside the Brilliant Minds of Google Founders Larry Page and Sergey Brin*. McGraw-Hill.

- Schwartz, S. E., & O'Brien, C. (2022). The Turritopsis dohrnii jellyfish: A model for biological immortality. Biology of Cells, 114(4), 285-292.

- Seim, I., Ma, S., & Gladyshev, V. N. (2014). Molecular signatures of natural selection in the beaked whale genome. Molecular Biology and Evolution, 31(4), 887-889.

- Shelley, M. (1818). Frankenstein. London: Lackington, Hughes, Harding, Mavor & Jones.

- Simmons, D. (2008). Epigenetic influences and disease. Nature Education, 1(1), 6.

- Sinclair, D. A., & Guarente, L. (2006). Small-molecule allosteric activators of sirtuins. *Annual Review of Pharmacology and Toxicology,* 46, 481-512.

- Sipp, D., et al. (2017). Clear up the confusion on stem cells. Nature, 545(7655), 19-21.

- Smith, J. (2003). The Quest for Eternal Life: Philosophical Perspectives. New York: Academic Press.

- Smith, J., & Chen, L. (2022). The future impact of genomic editing on aging. Journal of Gerontology, 77(2), 123-130.

- Smith, J., & Johnson, L. (2020). Generational dynamics in an aging society. Social Science Review, 34(2), 210-225.

- Smith, J., & Lee, R. (2021). The Impact of Remote Health Solutions on Healthcare Delivery. *HealthTech Journal, 16*(2), 112-129.

- Smith, J., Lee, R., & Patel, S. (2020). Ethical oversight in longevity research. Journal of Bioethics, 15(2), 123-135.

- Smith, J., Tang, M., & Yang, L. (2020). Designing for aging populations: Urban and architectural responses to demographic shifts. Journal of Urban Design and Planning.

- Smith, K., & Blanc, R. (2021). Cross-border Collaborations in European Longevity Initiatives. Journal of Gerontology Science, 76(1), 89-105.

- Smith, L. (2020). The cost of innovation: Balancing healthcare budgets in the age of advanced medicine. Medical Economics Review, 45(3), 23-34.

- Smith, L., Johnson, R., & Chen, T. (2021). Bridging the Gap: Making Longevity Accessible for All. Health Policy Journal, 45(2), 102-117.

- Smith, M., & Patel, R. (2018). Inequality in Health and Healthcare: Challenges and Solutions. World Health Organization.

- Smith, M., McNulty, H., & Eaton-Evans, J. (2020). Nutrition, vitamins, and minerals. Nutrition and Exercise Concerns of Middle Age. Clinical Sports Nutrition, 5(1), 83-95.

- Smith, R., & Johnson, L. (2022). The Genetic Frontier: CRISPR and Beyond. Cambridge Journal of Genetics.

- Smith, R., & Jones, L. (2020). Intellectual Property in the Age of Lifespan Extension. Biotech Law Review, 34(2), 210-225.

- Smith, R., et al. (2020). "The Economic Impact of Longevity Technologies". Economic Review, 89(3), 130-145.

- Social Security Administration (2021). The 2021 Annual Report of the Board of Trustees. Retrieved from [source].

- Sprent, J. I. (2001). Nodulation in Legumes. Royal Botanic Gardens, Kew.

- Suzuki, K., et al. (2018). Healthy aging practices from Japan: Insights and implications. Asian Health Review, 19(1), 11-19.

- Takahashi, K., Tanabe, K., Ohnuki, M., Narita, M., Ichisaka, T., Tomoda, K., & Yamanaka, S. (2007). Induction of pluripotent stem cells from adult human fibroblasts by defined factors. Cell, 131(5), 861–872.

Arlo Voss

• Takano, T., Nakamura, K., & Watanabe, M. (2002). Urban residential environments and senior citizens' longevity in megacity areas: The importance of walkable green spaces. Journal of Epidemiology and Community Health, 56(12), 913-918.

• Taylor, A. et al. (2021). Ethical Implications of Longevity Science. Journal of Bioethics.

• Taylor, P. (2020). The next America: Boomers, millennials, and the looming generational showdown. Public Affairs.

• Thomson, J. A., Itskovitz-Eldor, J., Shapiro, S. S., Waknitz, M. A., Swiergiel, J. J., Marshall, V. S., & Jones, J. M. (1998). Embryonic stem cell lines derived from human blastocysts. Science, 282(5391), 1145-1147.

• Thorpe, K. E., & Howard, D. H. (2006). The rise in spending among Medicare beneficiaries: the role of chronic disease prevalence and changes in treatment intensity. Health Affairs, 25(5), w378-w388.

• Tian, X., Azpurua, J., Hine, C., Vaidya, A., Myakishev-Rempel, M., Ablaeva, J., ... & Gorbunova, V. (2013). High-molecular-mass hyaluronan mediates the cancer resistance of the naked mole rat. *Nature, 499*(7458), 346-349.

• Timmermann, C., & Tulle, E. (2021). Theories of aging: historical perspectives. Cambridge University Press.

• Topol, E. (2019). Deep Medicine: How Artificial Intelligence Can Make Healthcare Human Again. Basic Books.

• Topol, E. J. (2015). The patient will see you now: The future of medicine is in your hands. Basic Books.

• Topol, E. J. (2019). Deep Medicine: How Artificial Intelligence Can Make Healthcare Human Again. Basic Books.

212

- Topol, E. J. (2019). High-performance medicine: the convergence of human and artificial intelligence. Nature Medicine, 25(1), 44-56.

- Topol, E. J. (2019). High-performance medicine: the convergence of human and artificial intelligence. Nature medicine, 25(1), 44-56.

- Van Parijs, P., & Vanderborght, Y. (2017). Basic Income: A Radical Proposal for a Free Society and a Sane Economy. Harvard University Press.

- Vayena, E., Blasimme, A., & Cohen, I. G. (2018). Machine Learning in Health Care: A Critical Appraisal of Challenges and Opportunities. eLife, 7, e32782.

- Vayena, E., Blasimme, A., & Cohen, I.G. (2018). Machine learning in health care: Ethical concerns for online research ethics committees. PLOS Medicine, 15(8), e1002689.

- Vincent, J. A. (2020). Old age: The Global Context. Routledge.

- Warburton, D. E., Nicol, C. W., & Bredin, S. S. (2006). Health benefits of physical activity: The evidence. CMAJ, 174(6), 801-809.

- Wilde, O. (1890). The Picture of Dorian Gray. London: Ward, Lock & Co.

- Willcox, D. C., Scapagnini, G., & Willcox, B. J. (2014). Healthy aging diets other than the Mediterranean: A focus on the Okinawan diet. Mechanisms of Ageing and Development, 136-137, 148-162.

- Williams, R. (2021). Environmental factors and aging in Australia. EcoHealth Journal, 26(5), 345-352.

- World Health Organization. (2007). Global age-friendly cities: A guide.

- World Health Organization. (2020). Physical activity. Retrieved from https://www.who.int/news-room/fact-sheets/detail/physical-activity.

- Wright, N. T. (2003). *The resurrection of the Son of God*. Fortress Press.

- Yalom, I. (2008). Staring at the Sun: Overcoming the Terror of Death. Jossey-Bass.

- Yang, G. (2018). The social significance of Confucian family values in aging. Journal of Family Values, 45(3), 231-251.

- de Grey, A. D. N. J. (2007). Life longevity escape velocity: Why genuine control of aging may be only 25 years away. *Impact Aging,* 9, 456-461.

- de Grey, A., & Rae, M. (2007). Ending Aging: The Rejuvenation Breakthroughs That Could Reverse Human Aging in Our Lifetime. St. Martin's Press.

www.ingramcontent.com/pod-product-compliance
Lightning Source LLC
Chambersburg PA
CBHW030004190526

45157CB00014B/422